The experiment of electricity production

The Secret of Electricity production

During my working life-time as electrical contractor and engineer I found that most of the electricians I meet will find hard to explain how the electricity is generated. Well established companies do not pay too much attention to the electricians training.IBEW is doing something but not enough.The point is : you need to be a self starter in order to achieve skills to protect yourself and others and be financially well.

There is no blame in this is just my intention help them a bit. I know how hard is to support a family, to bring food and things to our loves one since working hard.

It remains no time to review and to make theorethical practice in order to understand everything we do......unless we care about us !

For any question please e-mail me at:

axaelectric@yahoo.ca

Thank you for reading my book, and please take a minute and make your review. In case I disappointed you I will consider your point of view and make a revision for this book. Your review is important for me!

About The Author

Electrical Engineer

Low Voltage and 815V CC Electrical Room- Bucharest-Subway Station –electrical rehabilitation

Ajax Fire Station – Ontario, Canada- Main Electrical Room

All books with CORNEL BARBU as the **publisher**

EL LIBRO DEL ELECTRICISTA COMO LEER PLANOS ELÉCTRICOS (EL LIBRE DEL ELECTRICISTA) (Spanish Edition) (Kindle Edition) by CORNEL BARBU

ELECTRIC MOTORS-CONTROL DIAGRAM (SELF-STARTER UNIVERSITY) (Kindle Edition) by CORNEL BARBU

ELECTRICIAN'S BOOK -3-WAY&4-WAY LIGHTING SWITCH -WIRING DIAGRAM (SELF-STARTER UNIVERSITY) (Kindle Edition) by CORNEL BARBU

Electrician's Book -ELECTRICAL CONTRACTORS-12 STEPS ESTIMATING ELECTRICAL WORK (Kindle Edition) by cornel barbu

ELECTRICIAN'S BOOK -Fire Alarm System-Sketches & Diagrams (SELF-STARTER UNIVERSITY) (Kindle Edition) by CORNEL BARBU

Generating Electricity-Electric Field (Electrician's Book - Apprentice's Total Training) (Kindle Edition) by cornel barbu

Generating Electricity-Magnetic Field (Electrician's Book - Apprentice's Total Training) (Kindle Edition) by CORNEL BARBU

GUIDE for the Book : ELECTRICAL CONTRACTORS-12 STEPS- ESTIMATING ELECTRICAL WORK (Kindle Edition) by cornel barbu

POWER & LIGHTING SYSTEM BOXES-CONDUIT & WIRING CALCULATION (Kindle Edition) by CORNEL BARBU

Table Of Contents

Contents

Dedication
With love to

MIKE, JULIA & FLORIN

Lorenz Force

Conductive materials include a large amount of negative and positive charged particles. Subatomic particles exist in any matter or solids. Let take as example a coil with couple of turns made by copper. Lorentz force interacts with subatomic particles.

In order to see video demonstrations on your kindle device you'll need to download and install dolphin browser because it is the only one supporting flash-player. In case is difficult to do this suggest the reader use their computer to view them

Here is the link for video demonstration,

http://scratch.mit.edu/projects/12973067/#fullscreen

If this coil moves inside of magnetic environment, LORENZ FORCE acts on particles of electric charge due to a magnetic and the electric field (E&B).

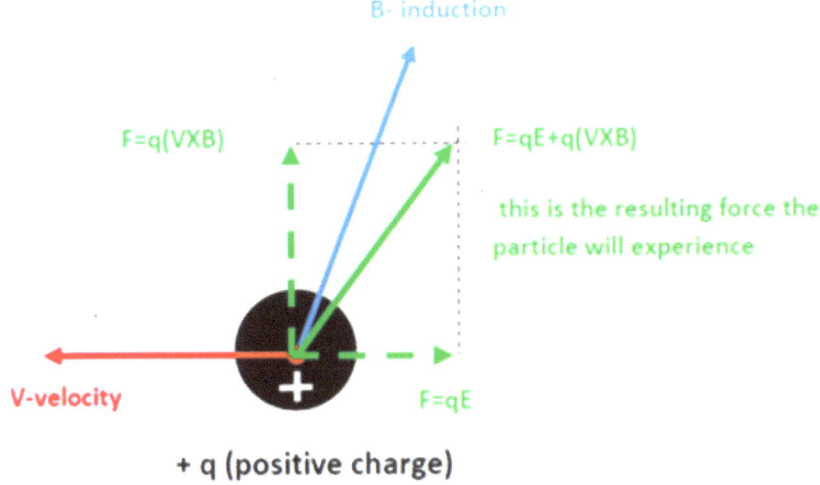

http://scratch.mit.edu/projects/13211447/#fullscreen

There is a formula to show the Lorentz Force (measured in Newton)

$$\mathbf{F} = q(\mathbf{E} + \mathbf{v} \times \mathbf{B})$$

Multiplying the q with the first and second term of equation will result:

F= q*E + q*VxB

There are two components of Lorenz Force: the electric force (q* E) and the magnetic force (q*VxB).Very important thing to keep in mind, the electric force is acting in the same direction the electric field is , since the magnetic force is perpendicular on the V and B vectors, up or down. The magnetic force is the means that "push" the electrons through the wire and assure a negative and positive charge accumulation at the extremity of the wire. This is equivalent with a voltage. This is the Electro-Motive-Force (EMF).
http://scratch.mit.edu/projects/12994084/#fullscreen

Since the motion of the conductor (**L**) <u>starts from point A</u> to B with the velocity V the positive and negative electric charges experiences LORENZ forces and moves in opposite directions. Some positive and negative charges rich the terminals of conductor and an electrical field appears (E1)

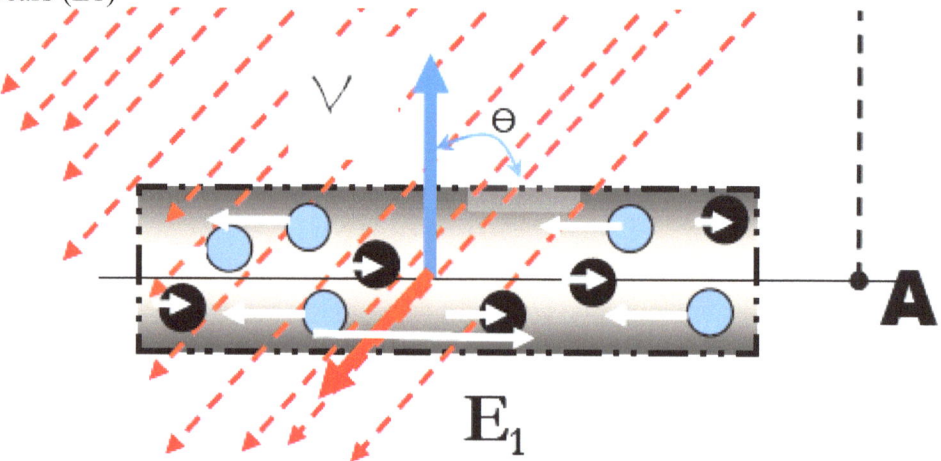

The motion continues with the same velocity V into the magnetic field Band more charges will accumulate. The electrical field strength will increase and become (E2)

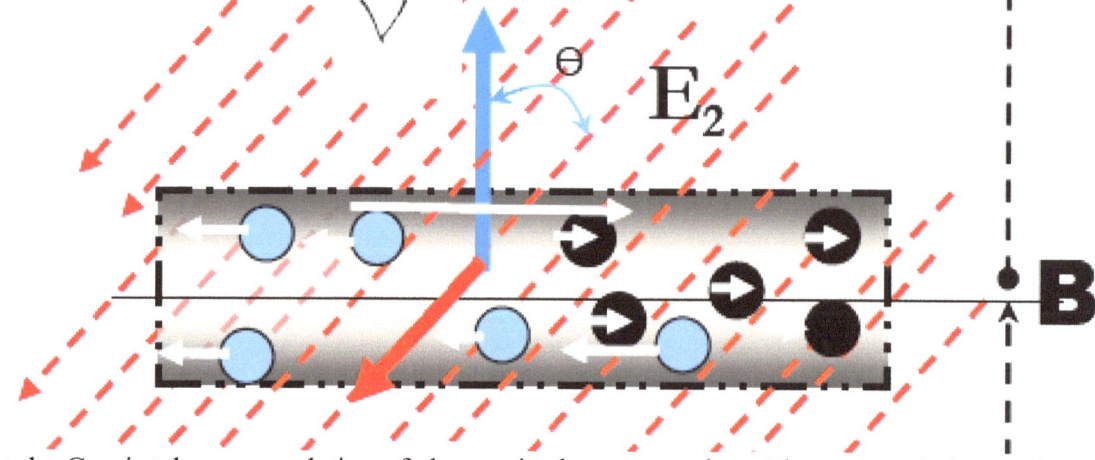

At the C point the accumulation of charges is almost complete. The accumulation will not continue indefinitely and will stop when the force created by the magnetic field (**B**) is equal with the force created by the electric field (E3).

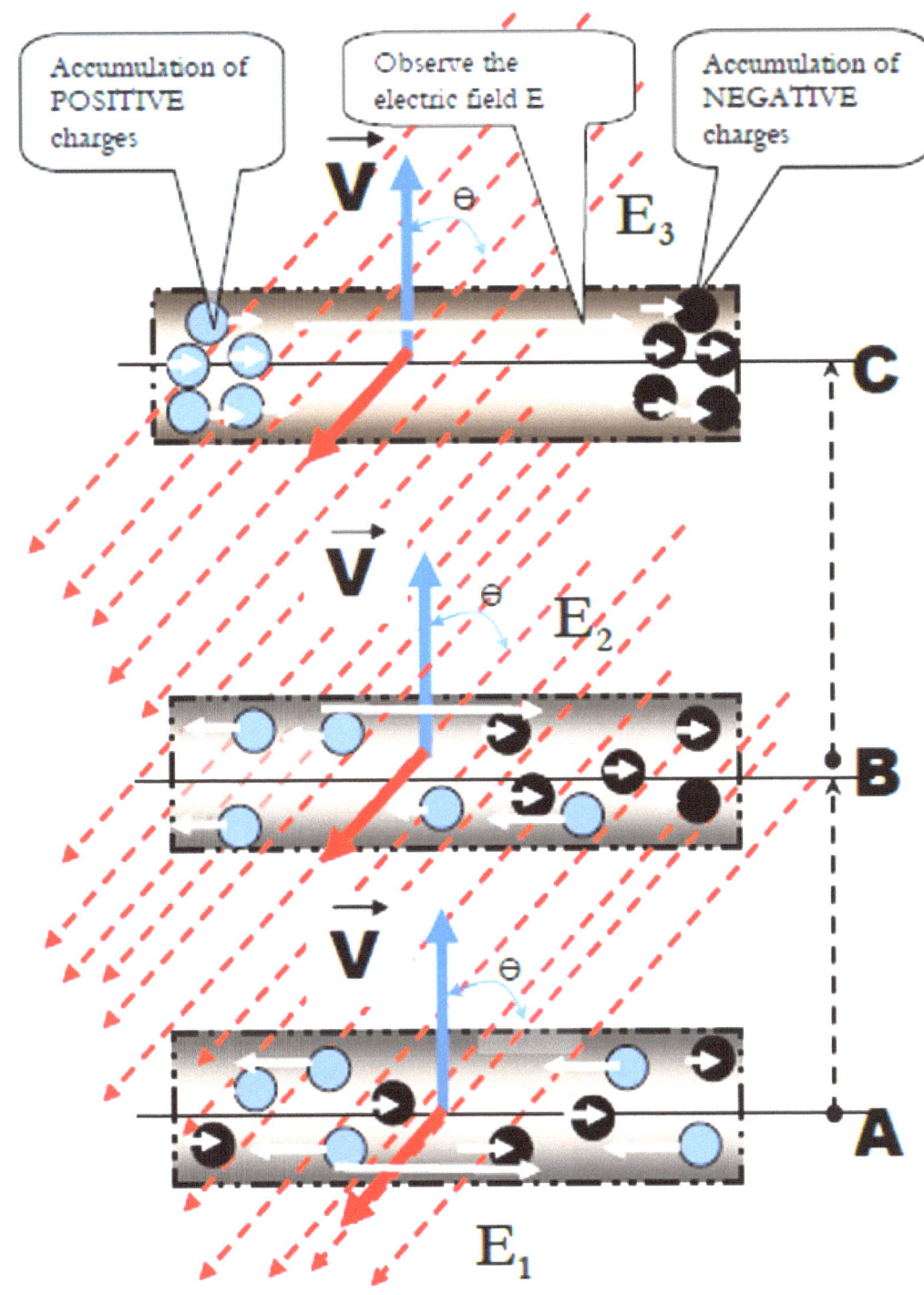

See my project showing the subatomic motion model live!
http://scratch.mit.edu/projects/13038510/#fullscreen

Faraday's Law

We learned from Michael Faraday the electromagnetic induction theory. Michael Faraday discovered that:

$$\mathcal{E} = -\frac{d\Phi_B}{dt}$$

 In other words make it easy to be understood by electricians, the circular motion of the coil will generate **a voltage** at the terminals of the coil and this voltage **is direct proportional** with the **ratio of how fast the magnetic flux changes in time**.

There are ways to generate VOLTAGE for a coil of wire (EMF-electro- motive- force). Conductive materials include a large amount of negative and positive charged particles. Subatomic particles exist in any matter or solids. Let take as example a coil with couple of turns made by copper that rotate inside of the magnetic field.

Lorentz force interacts with subatomic particles.

In order to see video demonstrations on your kindle device you'll need to download and install dolphin browser because it is the only one supporting flash-player. In case is difficult to do this suggest the reader use their computer to view them

$$\mathcal{E} = -\frac{d\Phi_B}{dt}$$

Here is the link for video demonstration,

http://scratch.mit.edu/projects/12973067/#fullscreen

If there is no motion will be no EMF.As per Faraday's law of induction the induced EMF in the wire is $\mathcal{E} = -\frac{d\Phi_B}{dt}$

"Electromotive force" is not considered a force, as force is measured in newtons, but a potential, or energy per unit of charge, measured in volts. Formally, EMF is classified as the external work expended per unit of charge to produce an electric potential difference across two open-circuited terminals" (

http://en.wikipedia.org/wiki/Electromotive_force

Permanent magnet and the electromotive force(EMF)

A magnet is any material or object that produces a magnetic field.

Magnetization is a characteristic of different objects. They hold magnetization for a long time Such material is a permanent magnet.

Next experiment shows what happens when such materials move around wires and vice versa when the wires move around the permanent magnet.

Actions like this determine a magnetic flux variation around the coil. In any situation we do the experiment will result a magnetic flux modification for the coil. As we mention before an electromotive force (EMF) is induce into the wire of the coil

For the next experiment we need to prepare an arrangement that includes:

- Bar magnet
- 6 turns coil made by copper
- Micro-voltmeter
- A resistive load
- 2-3 meters of copper wire

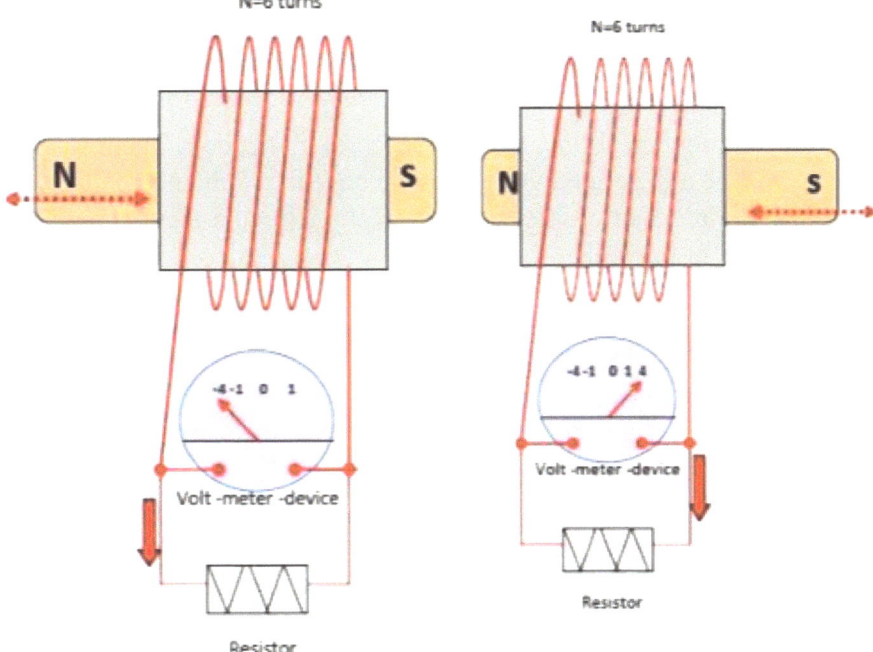

What need you to do is to move in and out the magnet bar inside the coil .The faster is done the higher voltage is indicated by the meter. As soon as you stop the motion the indicated voltage is zero.

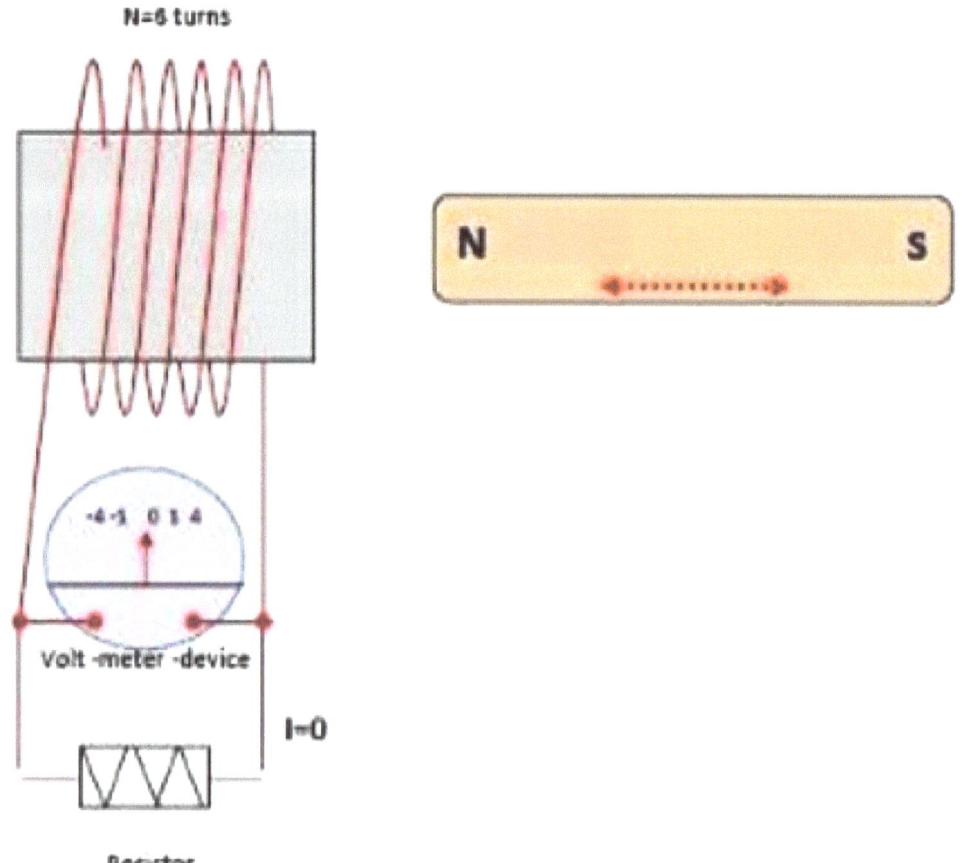

Please observe the Volt- meter indication when sliding the magnet through the coil. There is magnetic flux variation in time so will be a <u>generated voltage</u> at the coil; fact indicated by the volt-meter. The bottom line: Having no intersections between the coil's wire and the magnetic flux lines , will be no generated voltage at the coil.

I have prepared for you a special link I want you to see the "live experiment"
http://scratch.mit.edu/projects/13009608/ or
http://scratch.mit.edu/projects/13009608/#fullscreen

The amount of generated voltage (EMF) is directly proportional with numbers of turns of the coil. The number of turns is indicated by letter N.
The amount of generated voltage (EMF) is also directly proportional with the velocity of the magnet bar. The velocity is indicated by letter V.
Different coils with different numbers of turns located into magnetic field will generate different voltage values for each coil.
High velocity of the magnet bar generates high EMF induced in the coil
Do we need high voltage induced into the coil?
•More turns and high velocity will be required
Do we need low voltage induced into the coil?
•A reduced amount of turns and low velocity of the magnet bar will be required!

Do we need zero value for EMF?
• stop the motion of the magnet bar

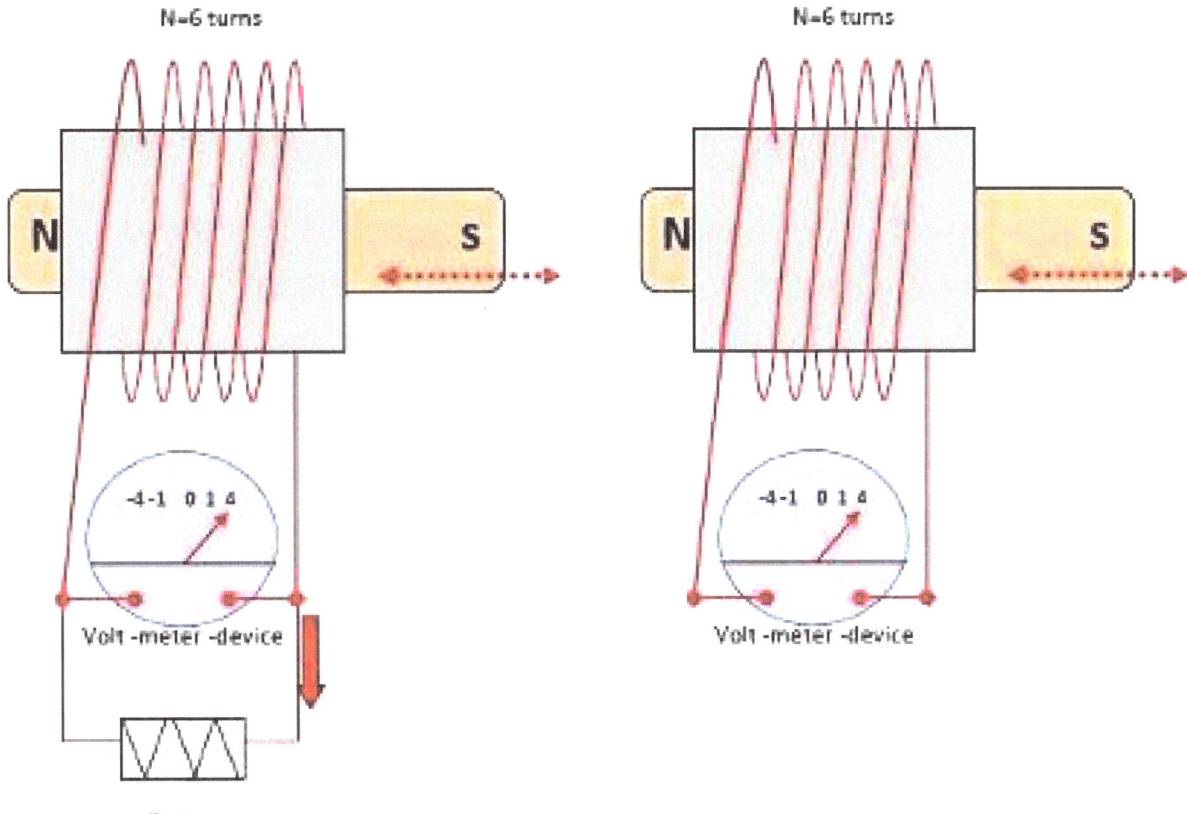

As per Faraday's Law when magnetic flux changes, is generated a voltage at the terminals of the coil. This voltage is measured by the volt-meter in both situations of electrical diagrams. In fact this is the induced electro- motive- force (EMF).If a resistor is connected at the coil's terminals as we have indicated in picture located at the left side, an induced current will flow through the resistor. Voltage is also at the coil's terminals. If the circuit is OPEN (with no resistor connected to the coil's terminals) only induced voltage as it's measured at the coil's terminals-see picture located at the right side.
Above picture is showing two different conditions:
the electrical circuit is closed by resistor(the load)
http://scratch.mit.edu/projects/13086199/#fullscreen

the electrical circuit is open
http://scratch.mit.edu/projects/13070395/#fullscreen

CONCLUSIONS:
According to Faraday's Law when magnetic flux changes in the proximity of a coil, is generated a voltage at the terminals of the coil. The magnetic flux changes when the magnet bar moves through the coil. The voltage is indicated by the volt-meter. In fact this is the induced electro-motive- force (EMF).This is a GENERATOR.....When a load(resistance) is connected to the wire, we have an induced current flows through the resistor. In this case a generator is producing electricity and the load is consuming this electricity. Voltage can be measured as well. If there is no load connected the circuit is OPEN but induced voltage is indicated by the voltmeter

Moving a coil into a magnetic field

Moving the coil inside the area where the magnetic field is present

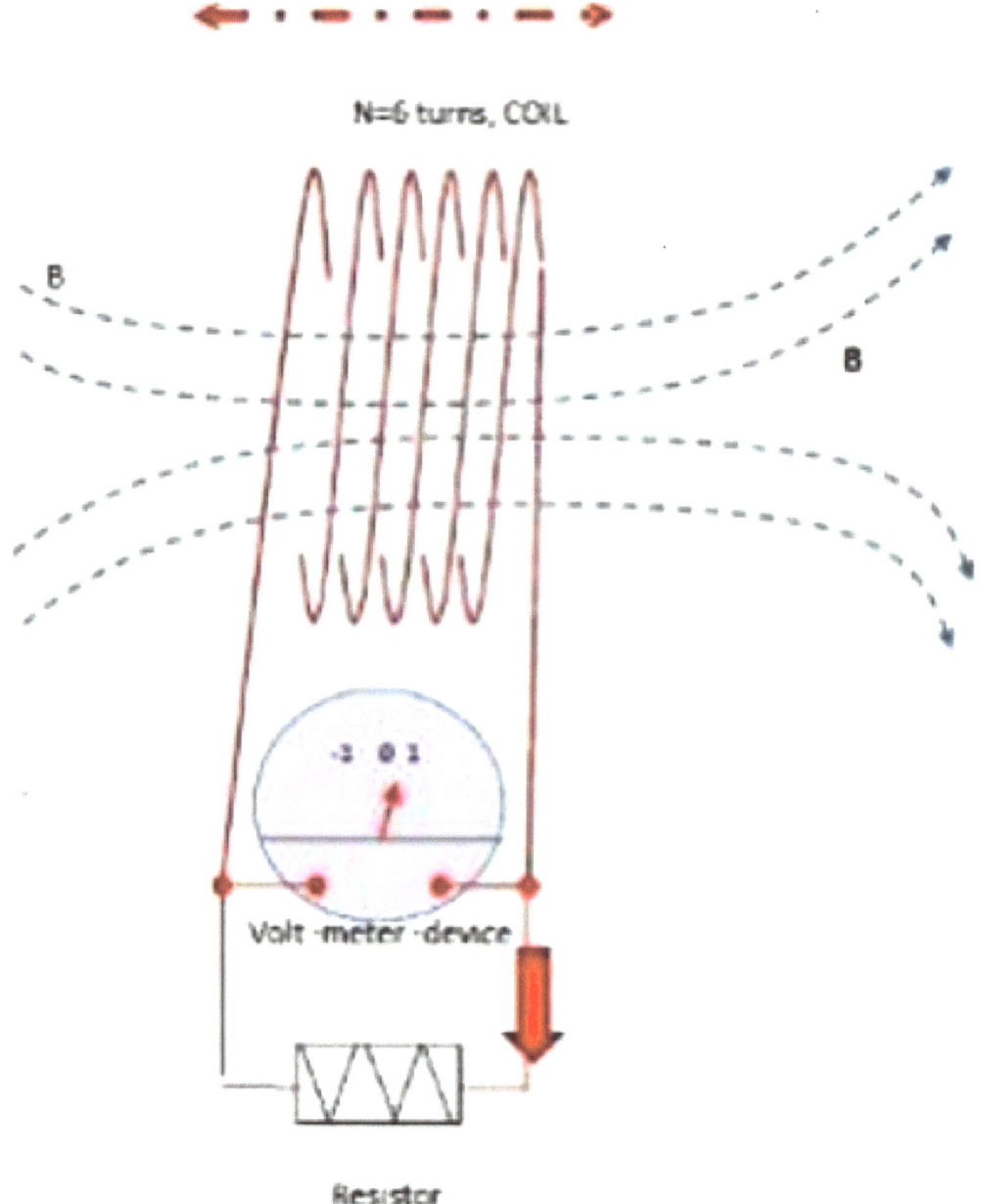

There is voltage generated at the terminals of coil as the voltmeter indicates. Moving the coil to the magnetic field with certain speed as per Faraday's Induction Law, an electro-motive –force named Voltage is generated at the coil's terminal. On the electrical circuit (coil and resistor) we observe the induced current. LIVE is bellow when you'll hit the links shown

http://scratch.mit.edu/projects/13925794/#fullscreen

http://scratch.mit.edu/projects/12995557/#fullscreen

Moving a magnet in front of a coil

The permanent magnet is moving in front of a coil. The magnetic field strength is B. The magnetic field induction is constant. The angular speed of the magnet is constant also.

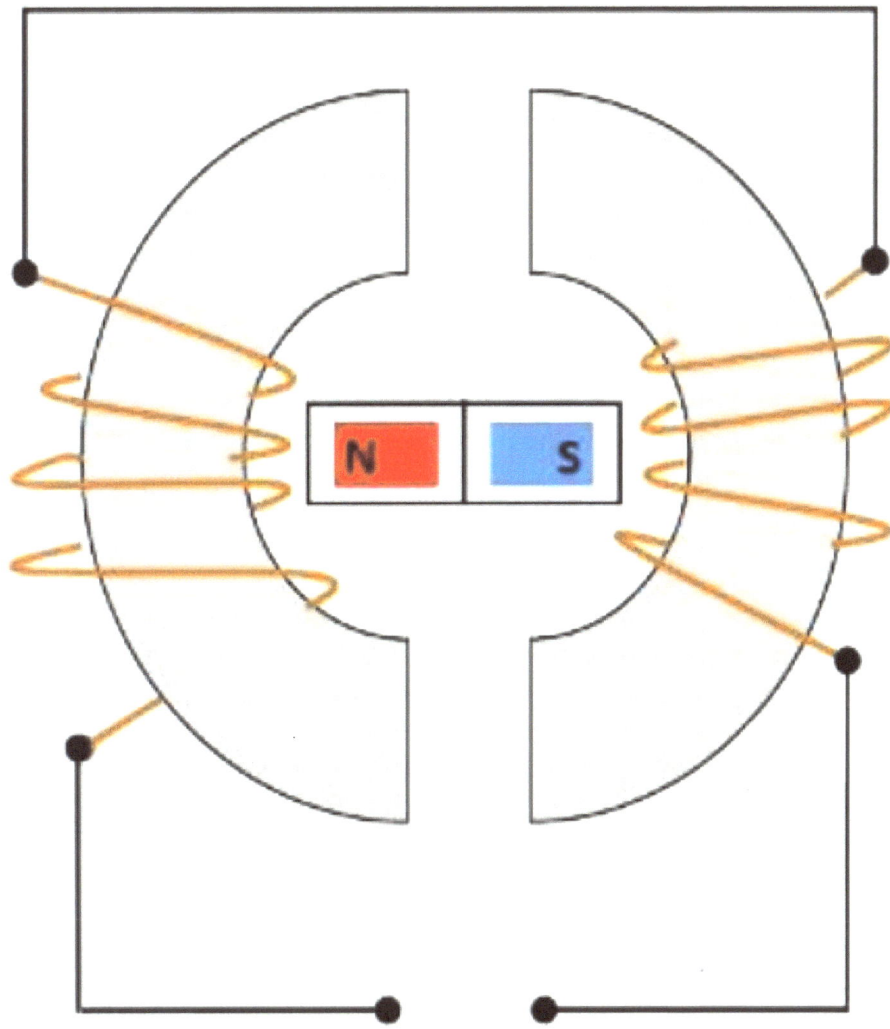

There is voltage generated at the terminals of coil as the voltmeter indicates. Moving the magnet in front of a coil, with certain speed as per Faraday's Induction Law, an electro-motive –force named Voltage is generated at the coil's terminal. On the electrical circuit we will observe the induced current in case the resistive load is connected to the coil's terminals... LIVE is bellow when you'll hit the links shown:

http://scratch.mit.edu/projects/13086199/#fullscreen
http://scratch.mit.edu/projects/13070395/#fullscreen
http://scratch.mit.edu/projects/13408595/#fullscreen

The Magnetic Flux

In order to find the definition of the magnetic flux (Φ) let me tell you that "flux" word is coming from Latin language where "fluxus" means "flow".

According to many dictionary the definition of magnetic flux is …

"The total magnetic induction lines crossing a surface": usually measured in webers or maxwells

Other definition could be:

"A measure of the quantity of magnetism, being the total number of magnetic lines of force passing through a specified area in a magnetic field."

In others word is the rate of flow per unit area. The permanent magnet creates magnetic field around them. Magnetic field is not visible but exists.

The magnetic field is in fact a field map.

1. When an electric charge is entering the magnetic field with velocity "V" will experience a force due to the strength (B) of the magnetic field. "V" is a vector."B" is a vector also. This force is the LORENZ force.

2. When a wire made by copper moves into the magnetic field with velocity "V" the free electrons of the material will experience the Lorentz force and under this force will start to move and to accumulate at the extremity of conductor.

http://scratch.mit.edu/projects/12994084/#fullscreen

http://scratch.mit.edu/projects/13316195/#fullscreen

http://scratch.mit.edu/projects/13038510/#fullscreen

3. When a coil made by copper is rotating into the magnetic field with velocity "V" the free electrons of the material will experience the Lorentz force also

The magnitude of the Lorenz force depends on:

•Amount of the electric charge (q)

•Magnetic field strength (B)

•Velocity (V) of charge moving into that environment

Is there a formula of the Lorentz Force? Yes it is ... and shows exactly what I mentioned at the above paragraph.

$F= q*v* B$

Magnetic Flux & Surfaces

We are going to study the rotation of a coil inside of magnetic field. The coil is a loop made by copper wire. The magnetic field intensity is constant and will not change in value (B=constant). Two permanent magnets will provide such magnetic field. The coil is rotating inside this magnetic field with angular velocity V

The scope of study is to observe if there is any interaction between the lines of the magnetic field and the coil.http://scratch.mit.edu/projects/13162517/#fullscreen

Changes of the magnetic flux for the coil that rotates- into the magnetic field

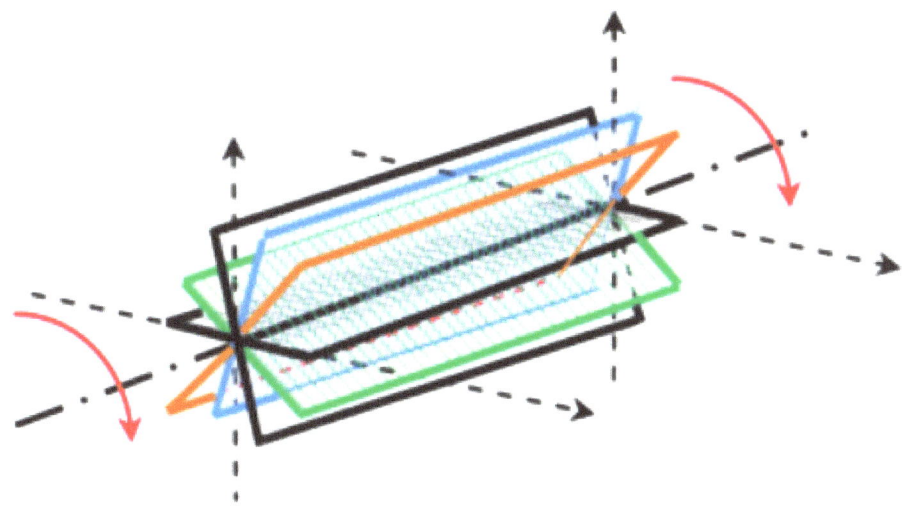

This is a very suggestive image how the coil is moving into the magnetic field, step by step. The rectangular areas in different colors show the position into magnetic field in 5 steps.

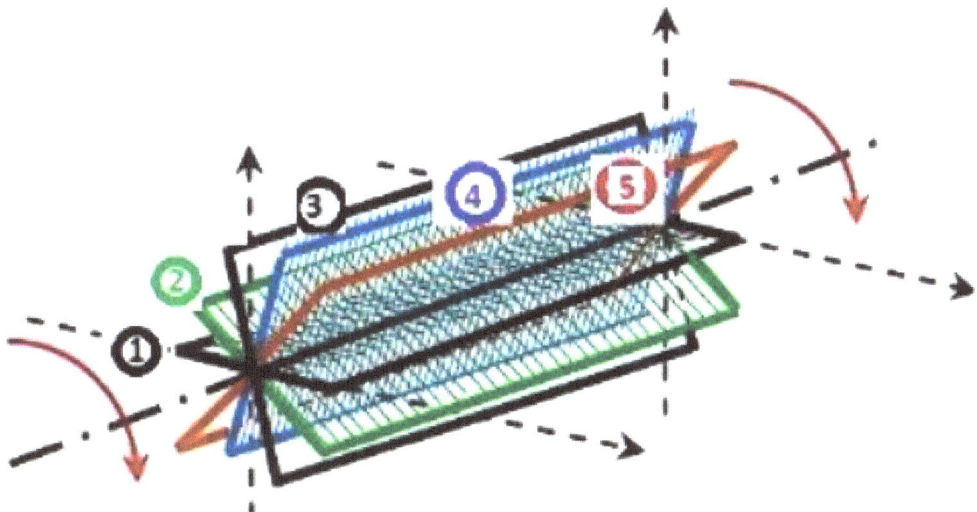

Red arrows indicate the clockwise rotation of the coil .The start point is the horizontal direction, indicated by BLACK area .After this will follow:

-Green position

-Black position

-Blue position and finally

-Red position

At each position the area of the coil will be punctured by certain amount of magnetic lines.For example at step 1 will be no magnetic field line to puncture the coil's area since the coil is parallel with the magnetic field line.

Since the coil will reach the position of step 2 magnetic field lines will puncture the area of the coil (see the green area)

When the coil will reach the position of step 3 more and more magnetic field lines will puncture the area of the coil (see the black area)

Since the coil will reach the position of step 4 maximum of the magnetic field lines will puncture the area of the coil (see the blue area)

When coil will reach the position of step 5 we observe decreasing of the magnetic field lines that puncture the area of the coil (see the red area)

Conclusion:

Because of the circular motion of the coil , the number of the magnetic field lines that puncture the area in motion will increase and decrease accordingly .This is equivalent with the magnetic flux change for the coil in time of rotation

The magnetic field strength B we declared to be constant (for example 2Tesla), The velocity of the coil we considered constant also (V=constant) . The area of the coil is constant also because we don't change the geometry of the coil.

The only things will change is the number of magnetic field lines will punctured the area of the coil at different step of motion. This is the magnetic flux that will change

http://scratch.mit.edu/projects/13208740/#fullscreen

Magnetic Flux and Voltage

In order to easy understand this magnetic flux change let see a front view of the coil and observe how will interact with different numbers of magnetic lines:

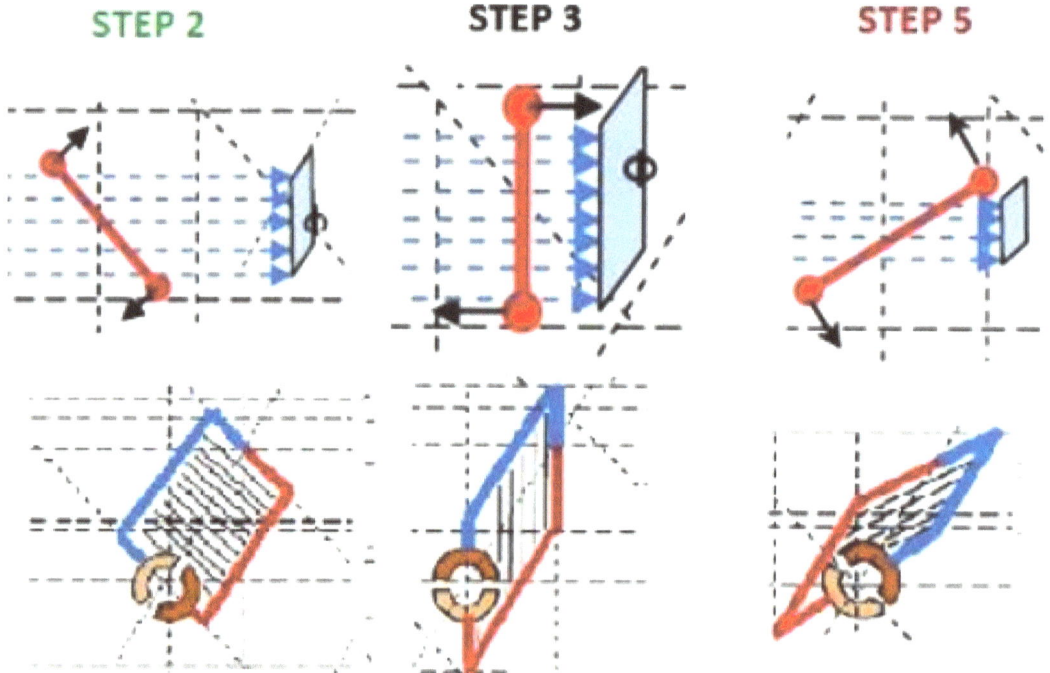

Blue area beside each position in step 2, 3&5 is highlighted to show the "quantity "of the magnetic flux. As we can observe **it is variable magnetic flux**.

The magnetic flux formula is shown below:

$$\Phi = B * A * Cos\ \Theta$$

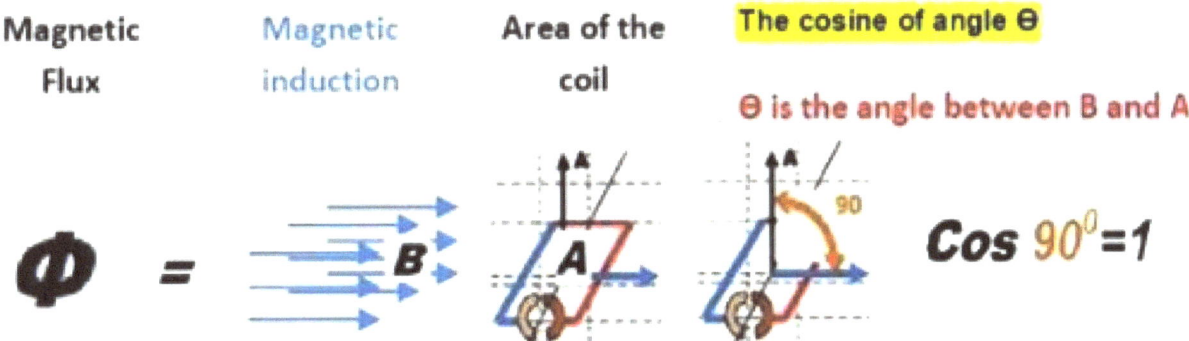

In other words different amount of magnetic lines will puncture the coil's area so the magnetic flux is changing due to the circular motion of the coil into this magnetic field. **This is a modification of the magnetic flux in time that will generate a voltage at the terminals of the coil.**

$$\varepsilon = -\frac{d\Phi_B}{dt}$$

The **minus sign** is the contribution of LENZ law: this means that the EMF the electromotive force (the voltage) is opposing to the flux change.

In mathematics this means that the EMF (electromotive force) is in fact a derivative function of the magnetic flux in time.

In other words make it easy to be understood by electricians, the circular motion of the coil will generate a voltage at the terminals of the coil and this **voltage is direct proportional with the ratio of how fast the magnetic flux changes in time.**

Electromotive force is measured in VOLTS and is in fact an electrical potential difference measured at the terminals of an open circuit.

For those with mathematics knowledge it is easy to understand the development of the EMF formula.

$$\xi = -\frac{d\Phi}{dt} = \frac{d(B*A*\cos\theta)}{dt} = -B*\frac{d(L*x*\cos\theta)}{dt} =$$

$$\xi = -B*L\frac{dx}{dt}(-\sin\theta) = B*L*V*\sin\theta$$

$$\xi = B*L*V*\sin\theta$$

Finally the most important thing to keep in mind is the last stage of the formula we see above:

$$\xi = B*L*V*\sin\theta$$

The EMF expression is indicating the parameters in charge to generate the voltage at the terminals of the copper coil that moves into magnetic field. Let's take one by one and explain them.

$$\xi = B*L*V*\sin\theta$$

The EMF expression is an equation. At the right side of the equation we identify the parameters in charge to generate the voltage at the terminals of the copper coil that moves into magnetic field. Let's take one by one and explain each member of the equation:

- B is the magnetic induction and measured in TESLA.
- Length of the coil is measured in meters and is indicated by letter L
- V is the angular velocity of the coil and is measured in m/s -1
- Sin θ – is the sin function of the angle θ. The angle θ is the angle between V and B vectors.

There is a direct proportional relation in between the EMF and these parameters. Any increase in value of induction for example will determine an increase in value for EMF. Similar things will happened for any increase of velocity or length of the coil.

In practice the length of the coil is a constant parameter and B and V also are going to be maintained constant during the motion of the coil inside of such magnetic field.

The only thing will change is the angle between the V and B vectors because of the circular motion.

http://scratch.mit.edu/projects/13162517/#fullscreen

This is important because periodically the angle between V and B is changing so……….. Sin θ does…….. So the EMF is changing also.

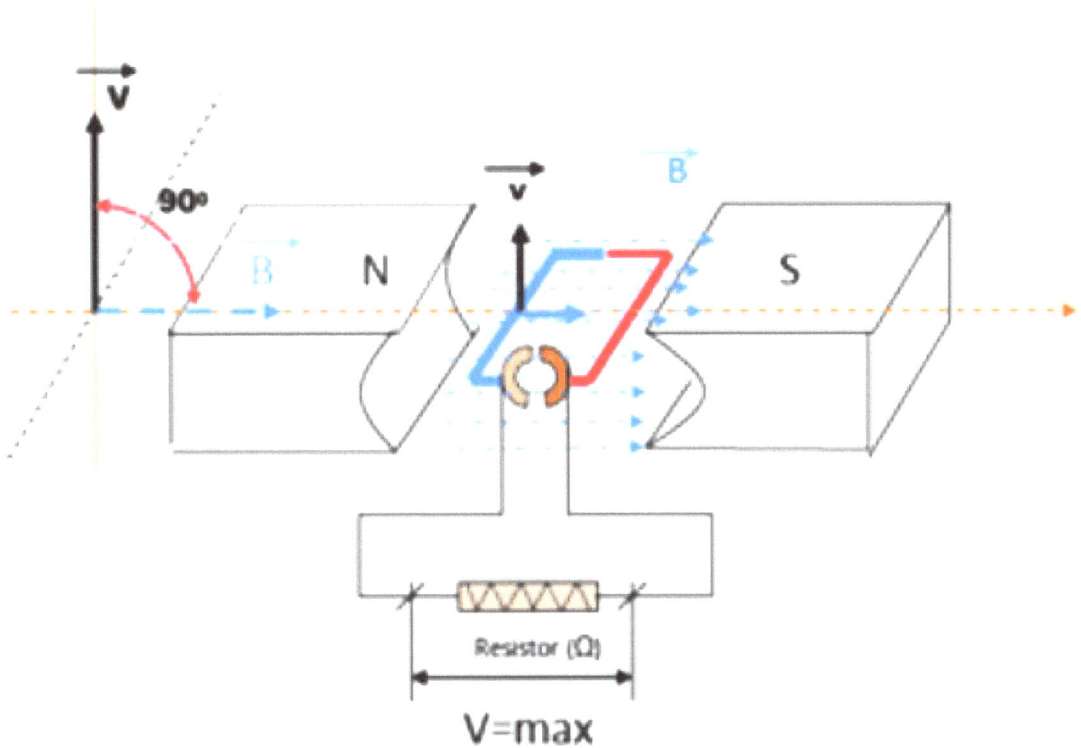

V=max

In this sketch we see 90 degree angle between the velocity and magnetic induction. According to the EMF formula in such position of the coil the EMF is going to have the maximum value since the *sin =1.*

EMF = L*B*V *1 =Max.

Magnetic flux variation

In the previous chapter at one point we find that: "This is a modification of the magnetic flux in time that will generate a voltage at the terminals of the coil". This voltage is EMF

$$\xi = -\frac{d\Phi}{dt}$$

There are two different ways to make possible the variation of the magnetic flux. One of the methods is been described before this page. That was the rotation of the coil with constant velocity inside of the magnetic field.

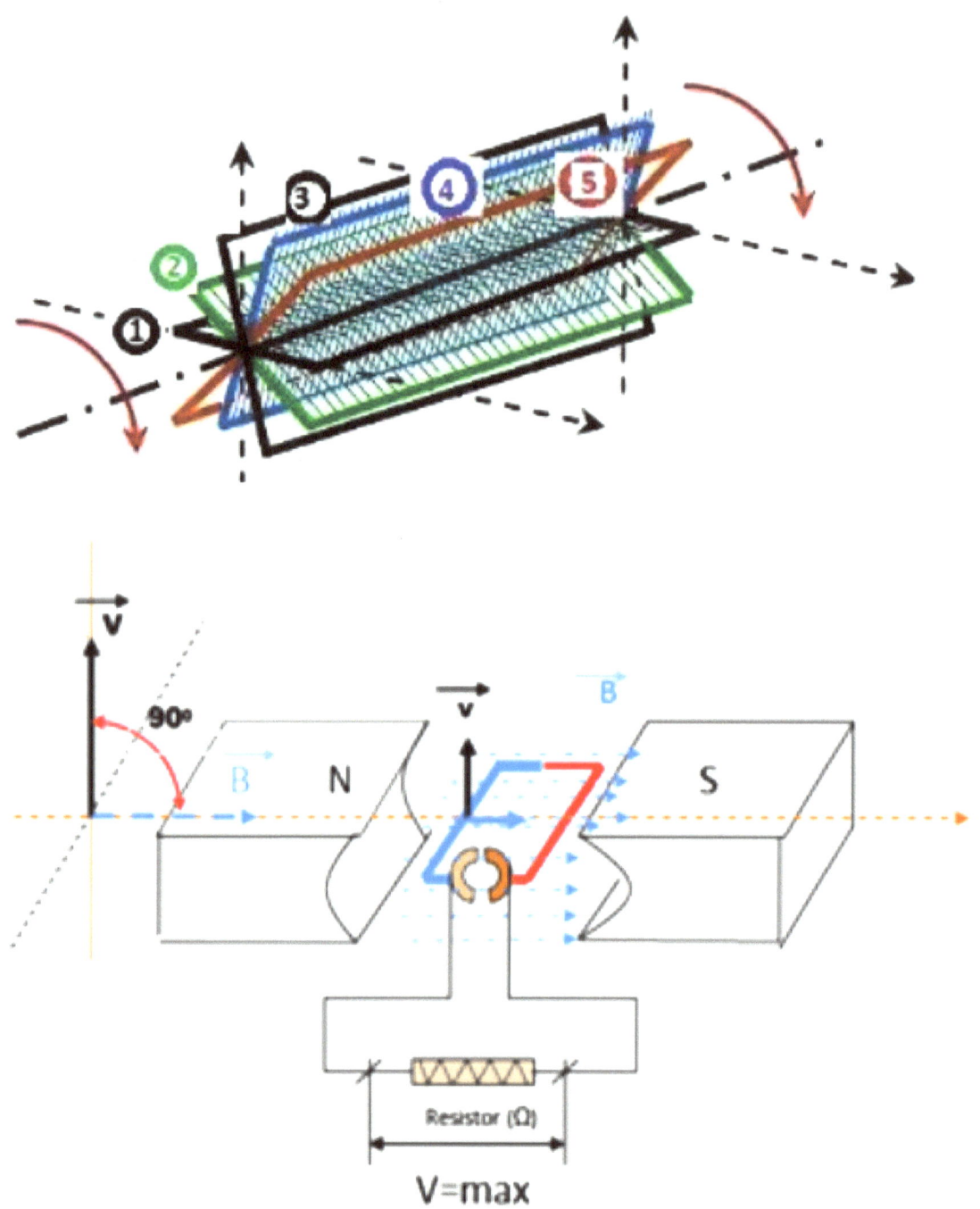

One other method is to rotate the magnetic field in front of the coil as indicate next sketch:

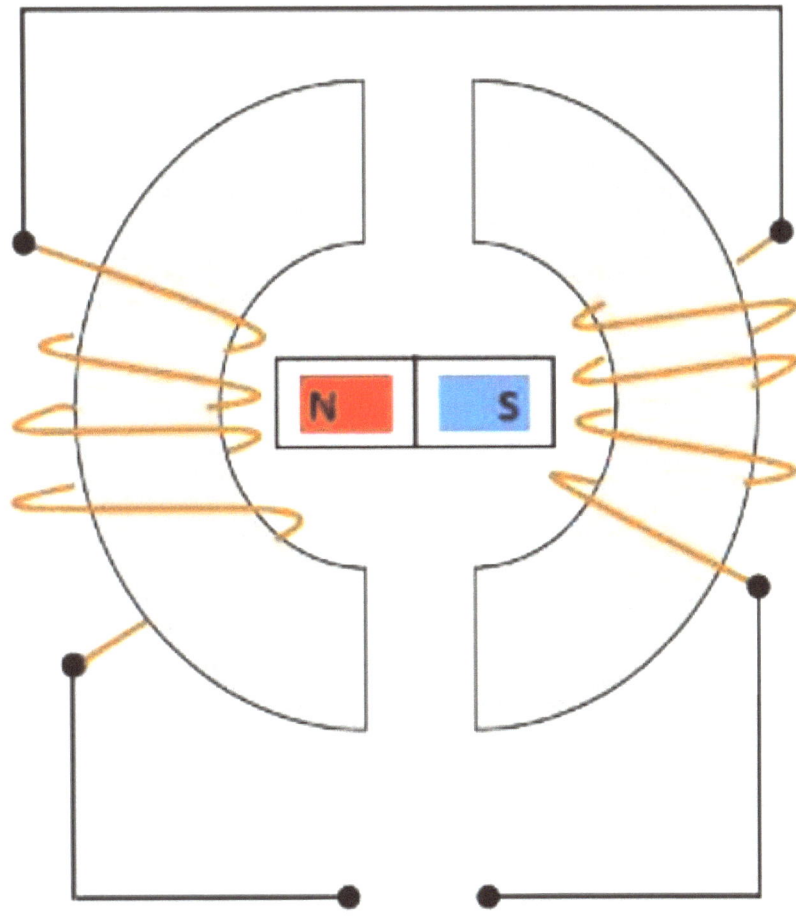

Either the rotation of the magnet in front of the coil or the rotation of the coil in magnetic field will generate EMF. It is required to mention one important thing. The magnetic field induction indicated by letter B is a constant parameter in both methods.

http://scratch.mit.edu/projects/13408595/#fullscreen

How the Electromotive Force will appear

"Electromotive force" is not considered a *force*, as force is measured in *newtons*, but a potential, or energy per unit of charge, measured in *volts*. Formally, EMF is classified as the external *work* expended per unit of *charge* to produce an *electric potential difference* across two open-circuited terminals" (

http://en.wikipedia.org/wiki/Electromotive_force)

http://scratch.mit.edu/projects/12973067/

The picture bellow is indicating a piece of copper conductor moving into the magnetic field with the velocity V from A to B and from B to C.

This is the angle between the vector **V** and **B** directions

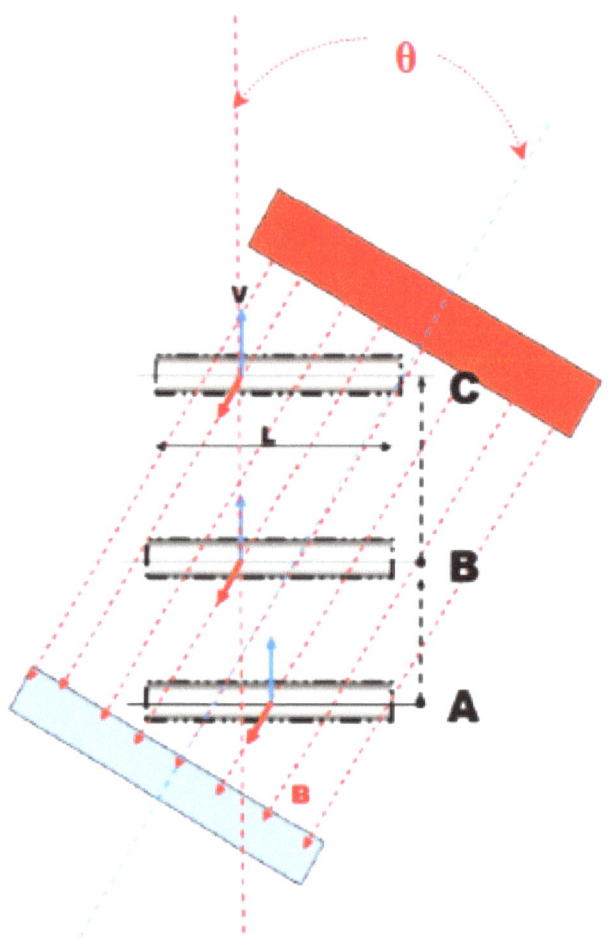

Moving the copper wire from A to B and from B to C the free electrons will polarize the extremity of the copper conductor and the electromotive voltage will appear .This voltage is direct proportional with the velocity V and the magnetic field strength B and most importantly with the Sinus of θ (where the θ angle is the angle between those two vectors: B and V)We "discover "the secret of electricity production! The angle between V and B being 90 degree or multiple of 90 degree then Sinus of θ is +1 or –1 so the voltage is reaching his maximum value. Let me explain you how this is happening…..Since the motion of the conductor (**L**) starts from point A to B with the velocity V the positive and negative electric charges experiences LORENZ forces and moves in opposite directions. Some positive and negative charges rich the terminals of conductor and an electrical field appears (E1)

http://scratch.mit.edu/projects/13316195/#fullscreen
http://scratch.mit.edu/projects/13852283/#fullscreen
http://scratch.mit.edu/projects/13162517/#fullscreen

Make no confusions in between electric field intensity (**E**) and the magnetic field intensity (**B**).

Electric field intensity (E)

Each positive ELECTRIC charge generates an **electric field around itself.** The ELECTRIC field lines around a positive charge have orientation AWAY from the charge and the same direction as the electrostatic force has against the "test charge".

- If the charge is POSITIVE the direction of the electric field is away to the electric charge.

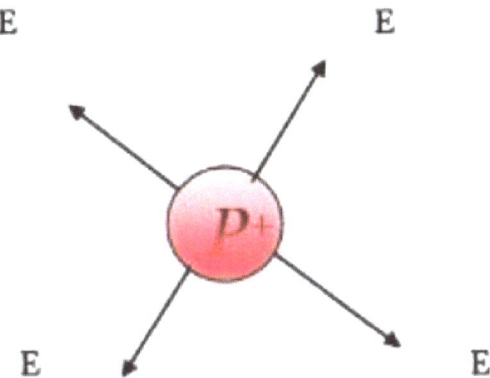

http://scratch.mit.edu/projects/15168575/#fullscreen

Each negative ELECTRIC charge generates an **electric field around itself** The ELECTRIC field lines around a negative charge have orientation TOWARDS the charge and the same direction as the electrostatic force has against the "test charge".

- If the charge is NEGATIVE the direction of the electric field is towards electric charge

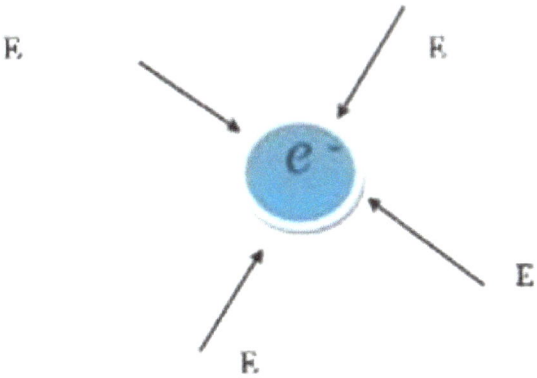

http://scratch.mit.edu/projects/15168825/#fullscreen

The electric field strength is indicated by the letter "E" and in fact is the magnitude of force per unit of charge and is measured in N/C (Newton/coulombs).

http://scratch.mit.edu/projects/15177736/#fullscreen

http://scratch.mit.edu/projects/15177645/#fullscreen

The electric field has a direction or orientation. The direction of the electric field is same as the electrostatic force the electric charge exerts to the test charge.

Magnetic field intensity (B)

The magnetic field is in fact a field map. The magnetic field is characterized by DIRECTION and STRENGHT also. Directions of magnetic lines are shown by arrows and the strength of the magnetic field by "B". B is a vector .When an electric charge is entering with velocity "V" into magnetic field will experience a force   due to the strength of the magnetic field. "V" is a vector also. This force is the LORENZ force.)

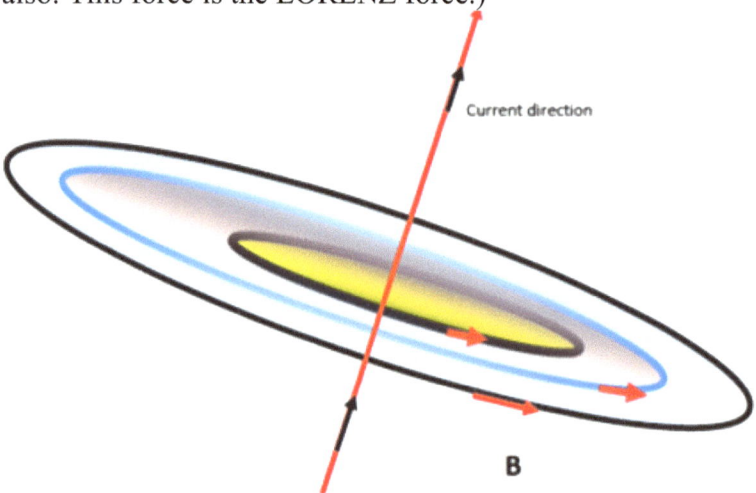

This is how magnetic field map is look like around a wire conductor where current flows…

http://scratch.mit.edu/projects/15178406/#fullscreen

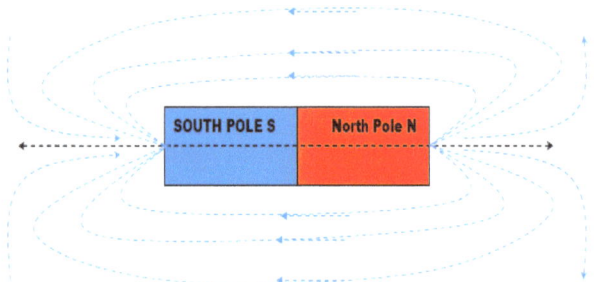

This is how magnetic field map is look like around magnets. Please observe the magnetic field lines……… The next link describes the paragraph where we analyze the copper conductor motion into magnetic field (B) with velocity V

http://scratch.mit.edu/projects/16034545/#fullscreen

So….. Coming back to our idea again let analyze the situation we have when moving an cooper conductor into the Magnetic field with a specific velocity:

Since the motion of the conductor (**L**) <u>starts from point A</u> to B with the velocity V the positive and negative electric charges experiences LORENZ forces and moves in opposite directions. Some positive and negative charges arrive at the terminals of conductor. Once the charges will begin to accumulate at the extremity of the wire an electrical field appears (E1)

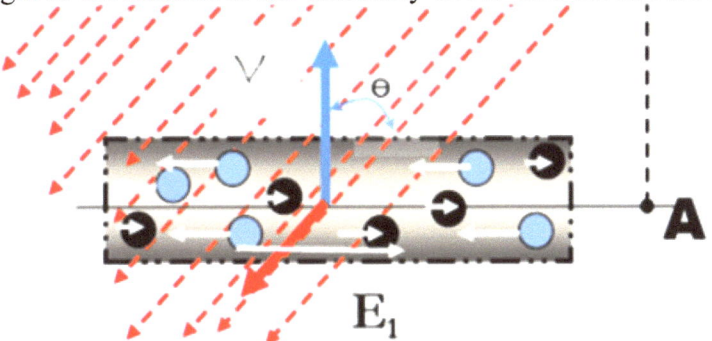

The motion of the wire continues with the same velocity V into the magnetic field B and more charges will accumulate. The electrical field strength will increase and become (E2)

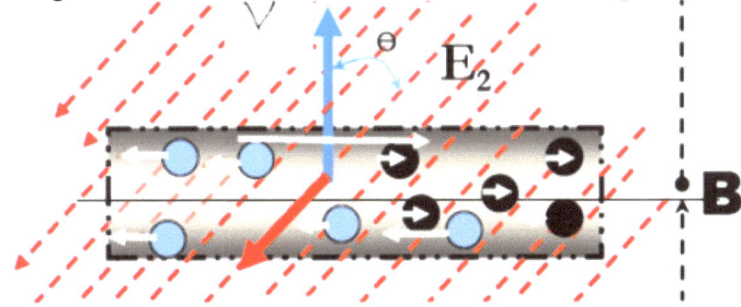

At the C point the accumulation of charges is almost complete. The accumulation will not continue indefinitely and will stop when the Lorentz force "manipulating" the charges is equal with the force created by the electric field (E3).

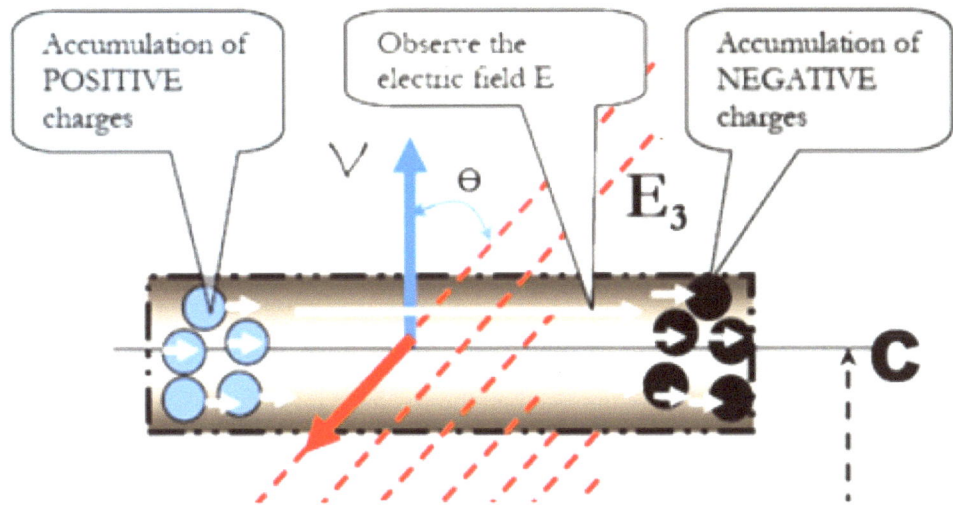

At this picture you'll see exactly what I mentioned before: THE CHARGES ACCUMULATION at the extremity of the copper wire. One end is "full" with **positive** charges (since the other end with **negative** charges (electrons)

The accumulation of charges is equivalent with a voltage .

This is EMF (electromotive force). It is measured in volts (v)

The EMF formula is:

$$U= V*B*L*sin\Theta$$

Wire polarization with electric charges

I will enter a little bit deeper into this matter to make sure you'll understand how this travel of the electric charges is happening because of the Lorentz Force.
We know that matter is made up of atoms and molecules. When there are actions of Lorentz force against electrons we might be in two situations:
•   Some atom loses electrons will have a negative charges deficit so will have more positive charges in excess. This atom become a CATION.
•   Some the atom gain electrons will have negative charges in excess so will have positive charges as deficit. This atom become a ANION

This words CATION&ANION is been introduced by Michael Faraday

This is in fact the accumulation of the Anions(- - - - - -) and Cations (+ + + +) to the terminals of the wire."The accumulation of cations and anions will not continue indefinitely and will stop when the force created by the magnetic field (**B**) is equal with the force created by the electric field (Ef)."

To make it simple let say accumulation of negative and positive charges at the extremity of the copper wire.

http://scratch.mit.edu/projects/12973067/#fullscreen

.http://scratch.mit.edu/projects/13639971/#fullscreen

"The accumulation will not continue indefinitely and will stop when the force created by the magnetic field (**B**) is equal with the force created by the electric field (E3)."

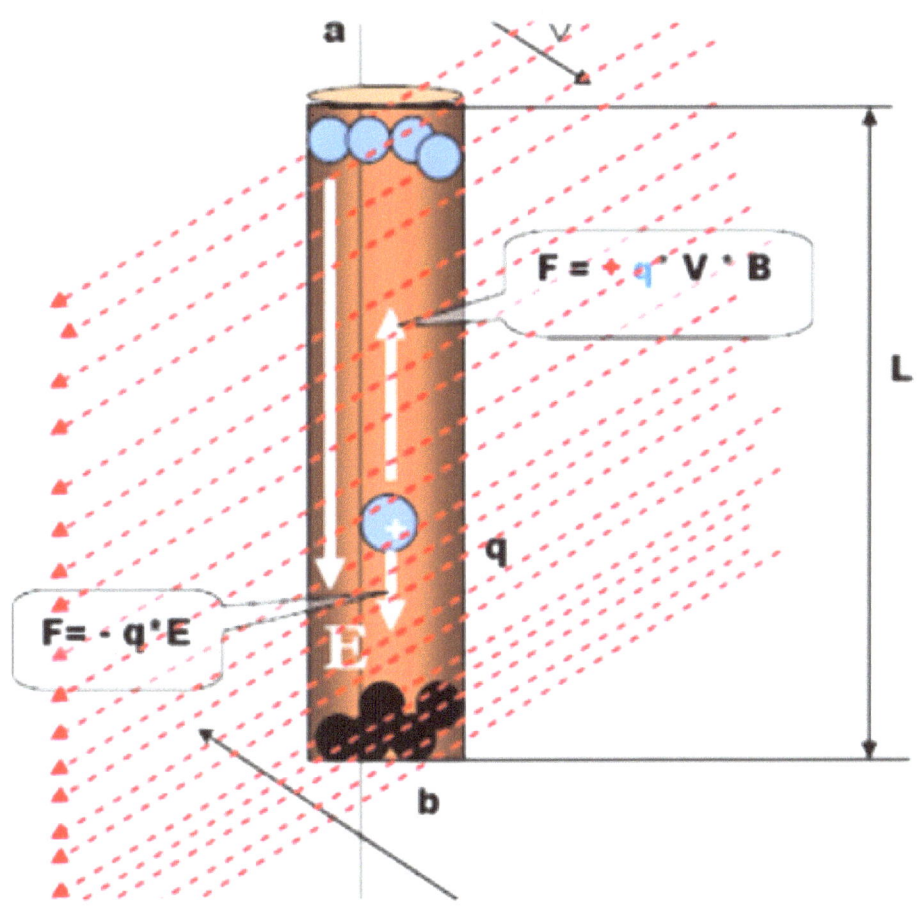

$$\vec{F}_{UP} = + q^* \vec{V} \cdot \vec{B} \text{ (+ indicates force is UP)}$$

$$\vec{F}_{down} = - q^* \vec{E} \text{ (- indicates force is down)}$$

Motion stops when $\quad -q^*\vec{E} = q \cdot \vec{V} \cdot \vec{B}$

Eliminating q from this equation will result: $\vec{E} = \vec{V} \cdot \vec{B}$

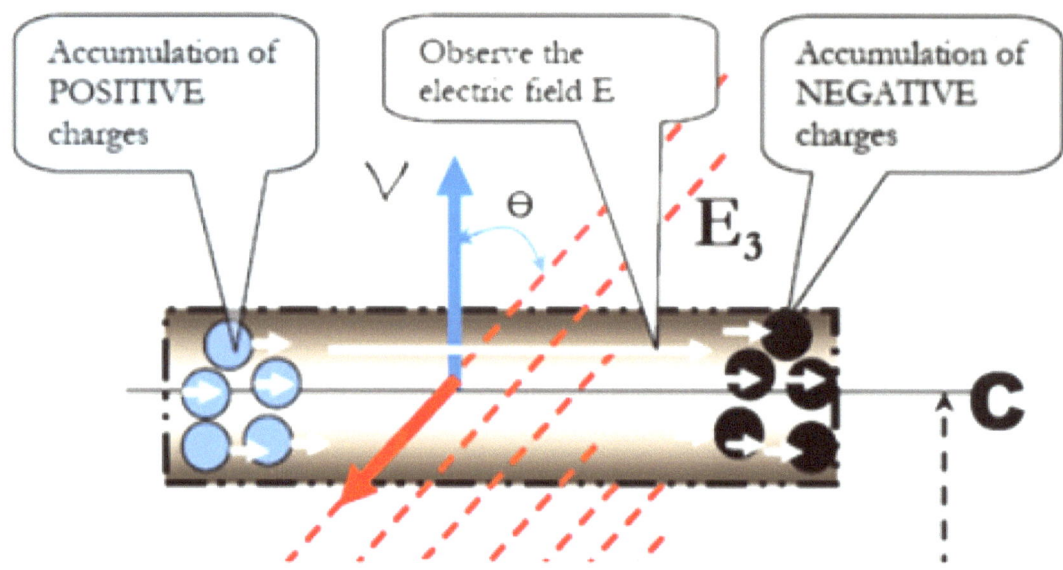

At this point (C) in space and time because of the accumulation of the charges at the termination of the conductor there is induced electromotive force (ξ) **that is equivalent with <u>a voltage</u>.** That is a source of energy.

EMF formula: U=V*B*L*sin θ

Please try to understand that each word I mention has his importance. In order to observe the importance of word each time I highlighted the word.

At this stage we "discovered" the VOLTAGE word

This is very important step where we find that a wire moving into the magnetic field (B) with a specific velocity (V) will experience a force that will direct the free electrons to the wires extremities .One end will accumulate the positive charges since the other end will accumulate the negative charges. The charges are moved over there because of the Lorenz force . At one end will be positive spot and other negative spot .This will create a difference of potential known as VOLTAGE. As soon as we connect the end of such conductor to a load we are going to identify an electrical current flowing through the load. The measuring device is indicating the existence of the electromotive force: the voltage as you can see bellow…The formula of the voltage (U) is showing also how this voltage depends on:

1.   Velocity of conductor

2.   The strength of the magnetic field

3.   The conductor's length and

4.   The angle of moving into the magnetic field.

How this angle is affecting the voltage value will see later on the next pages……..

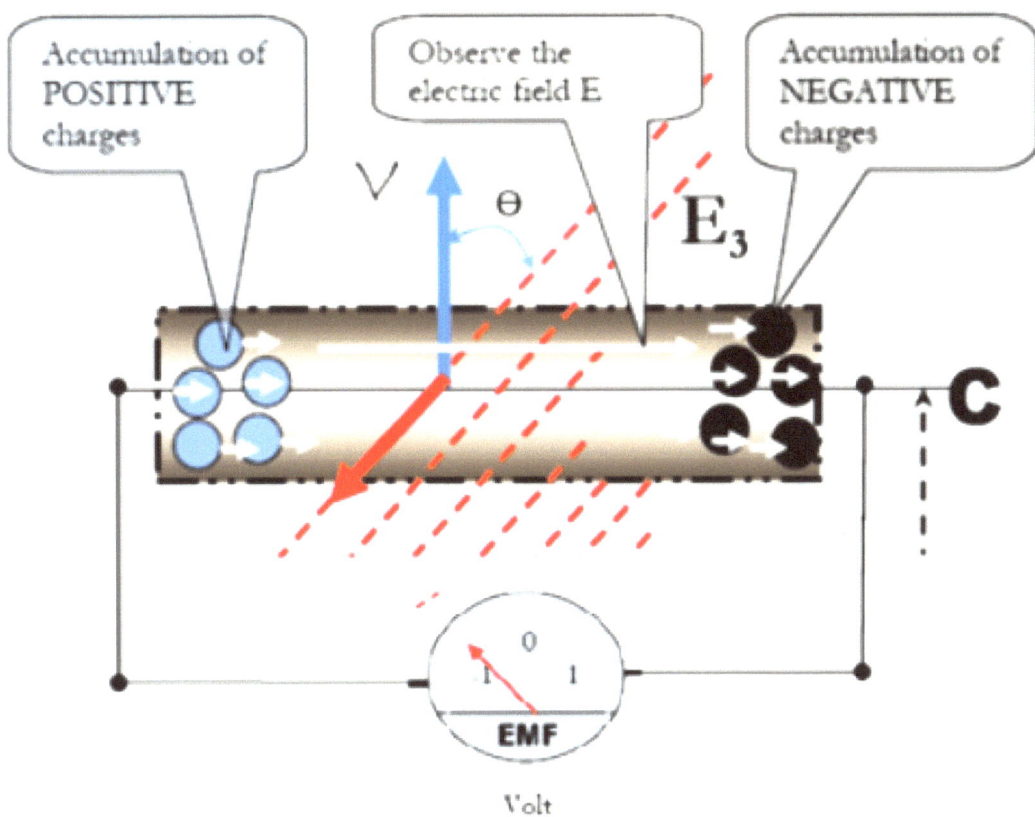

$$U = V * B * L * \sin \theta$$

At this point we agree that we found something crucial important:

1. Magnetic field

2. Conductor

3. Velocity

4. Voltage = $V * B * L * \sin \theta$

Rotating a coil in magnetic field

There are more ways to generate the electromotive force(EMF). EMF is in fact a voltage.To do our experiment we need to have available magnetic field, and a coil made by copper wire.Magnetic field is going to be supply by a pair of permanent magnets. Intensity of magnetic field is a measurable parameter and most of the time is indicate by letter B.Magnetic field intensity (B) is measured in Tesla (T).This was done as appreciation to Nikolai Tesla, the inventor.

This parameter will " tell us" how strong is this magnetic field.We consider B constant in value. The coil is moving with a constant velocity V.Since the coil is moving the angle between the velocity vector and B vector is changing .

The velocity is represented by a vector . The magnetic field intensity is represented by a vector also. We need to use this type of representation in order to understand the relation between these two parameters.The next picture will show you what exactly these vectors are and what is the relation between :

Velocity(V)

Induction (B)

Voltage (EMF)

The velocity is a vector .The next picture will show you how exactly this vector is represented:

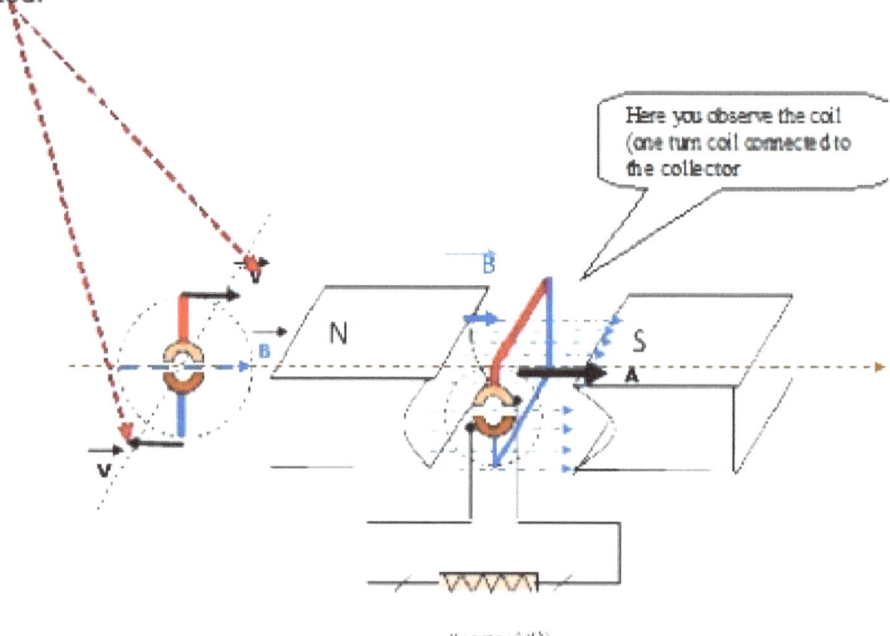

Here you observe the coil (one turn coil connected to the collector

There is something that needs clarification because we never discussed about it. On the picture we see above , where the coil is in between the magnet poles N&S observe the black arrow indicated by letter **"A".** This is the way is represented the area of the coil. Why is important to consider this new "item"? Well …because if we pay attention to details we are going to observe that this is the surface "washed" or "punctured „by the magnetic field lines indicated by B. Because of the rotation, the loop of coil will be in different position relative to the magnetic field lines. Less or more magnetic field lines will "wash" the coil's area (A) .Here are shown two positions of the coil where different area of coil is washed by magnetic field lines:

100% of the coil's area is punctured by the magnetic field lines in this position of the coil (observe Velocity vector parallel with B vector)

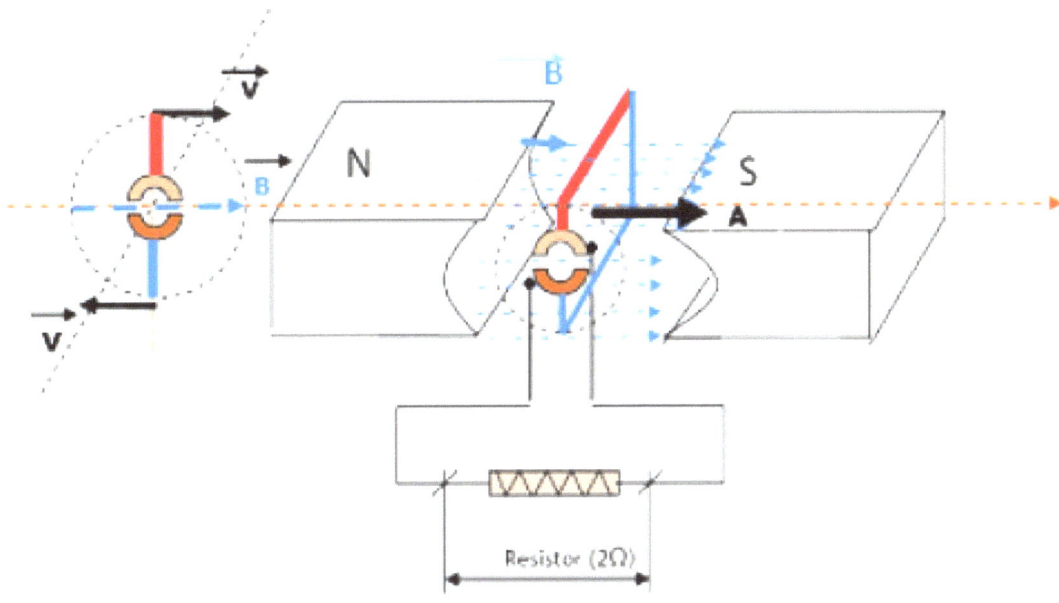

Looking to the next sketch we observe that 0 % of the coil's area is punctured by the magnetic field lines (observe Velocity vector perpendicular on B vector)

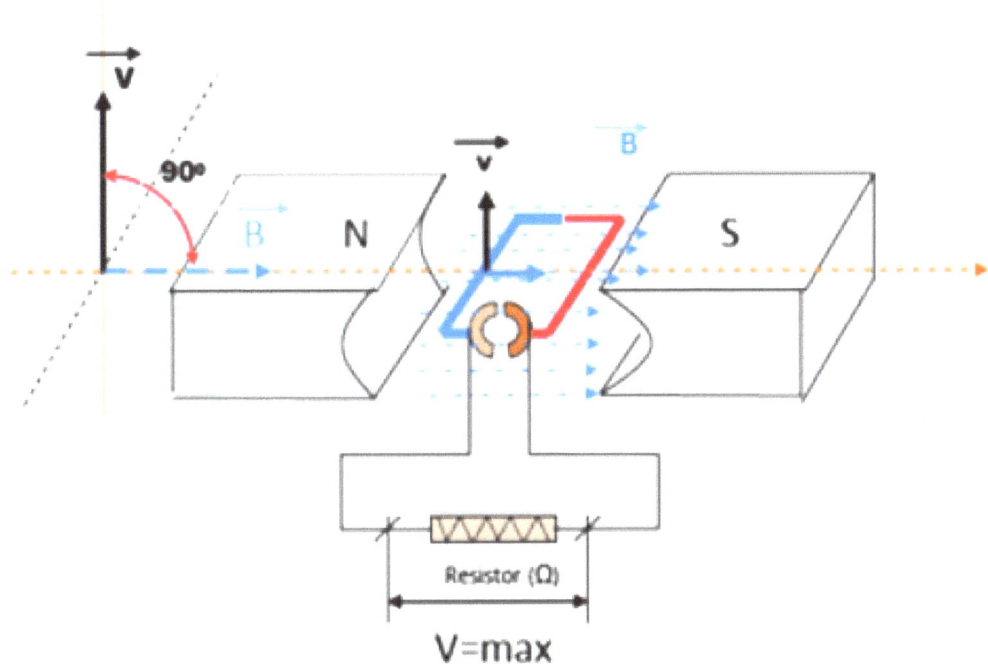

http://scratch.mit.edu/projects/13162517/#fullscreen

The above picture displays circular motion in three steps:
0 degree
90 degree &

180 degree

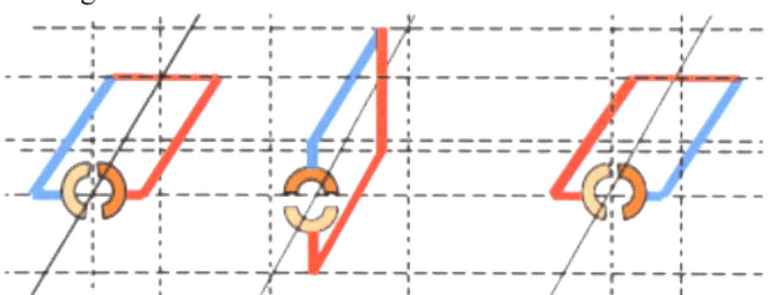

The motion continues displaying the position corresponding to:
270 degree
360 degree

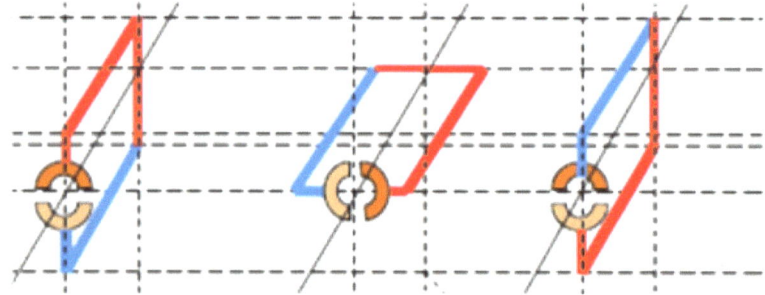

Observe the velocity vector V on the front view, just below the 3D view coil.

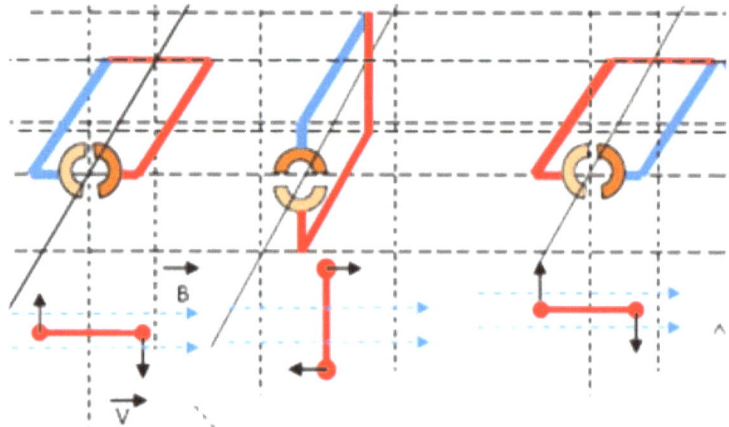

Zero degree V&B   /   90 degree V&B   /   180 degree V&B

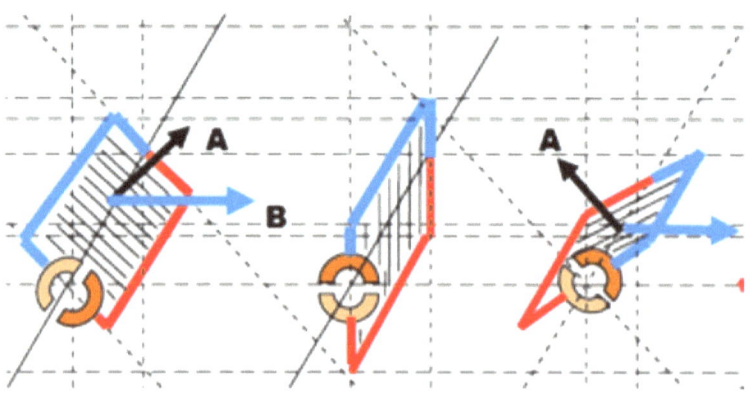

Variable area of the coil is punctured by the magnetic field lines.This is equivalent with variable magnetic flux. http://scratch.mit.edu/projects/13162517/#fullscreen

The experiment of electricity production

http://scratch.mit.edu/projects/12778668/

During my working life-time as electrical contractor and engineer I found that most of the electricians I meet will find hard to explain how the electricity is generated. Well established companies do not pay too much attention to the electricians training.

IBEW is doing something but not enough.The point is : you need to be a self starter in order to achieve skills to protect yourself and others and be financially well.

There is no blame in this is just my intention help them a bit. I know how hard is to support a family, to bring food and things to our loves one since working hard.

It remains no time to review and to make theorethical practice in order to understand everything we do......unless we care about us !

It remains no time available to review the books and notes we keep. Time to time we need to make theorethical judgements and review our notes so we understand everything we face day by day since workingThe Electricity Poduction is well described by the next experiment we are going to perform .In fact we are going to apply and understand the applications of FARADAY's law.In order to understand production of electricity we need to find answers to questions like:

What is the alternating current system.

What is the VOLTAGE and what is the electrical CURRENT

The system is describes the relation between the current and voltage.

How, why and when the voltage appears ?

There is a form or expression we represent such parameter?

Same questions we need to have for the current also...........

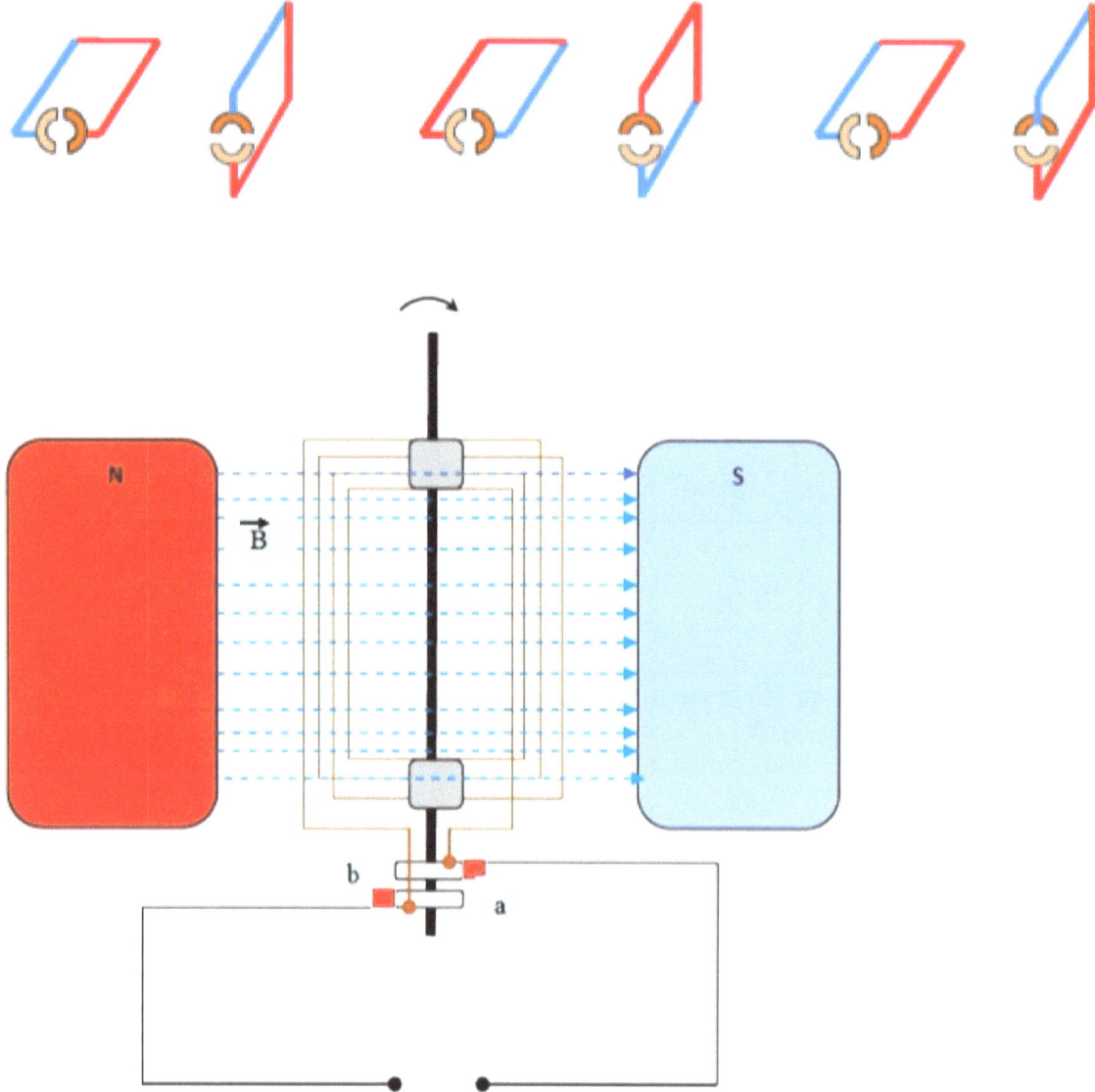

Rectangular wire loop rotating at angular velocity ω in radially outward pointing magnetic field B of fixed magnitude.

In the above picture we have the schematics of the **generator.**

This kind of <u>generator</u> is *producing electricity*. As you can see we have a **magnetic field** created by two magnets (N and S) or electromagnets. Also in between the magnets will rotate (spin) a *4 turns coil*.

The coil is attached to the central rod (the black line) and will have the wires connected to the *collector's part: a & b*. All these together: the rod, the coil and the collectors **a & b** is in fact the parts in circular motion: the ROTOR of the generator.

The magnet N&S is the motionless part: the STATOR of the generator.

What we have in fact is a device that will include a coil to rotate inside the magnetic field.

The intensity of **the magnetic field (B)** is the strength of the magnetic field. The magnetic field lines have a direction, from North Pole (N) to South Pole(S). In order to indicate the direction of the magnetic field B letter is used as a vector. The coil is rotating clockwise as been indicated by the black arrow on the drawing. The next view will show you the coil rotation .Observe the blue line to understand the clockwise rotation of the ROTOR into the magnetic field. Next link shows this..

http://scratch.mit.edu/projects/13162517/#fullscreen

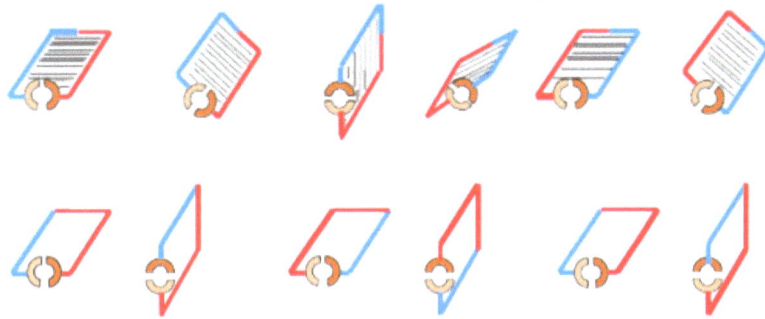

Having this installation ready for our experiment let analyze the response of this interaction between the magnetic field and the copper coil. By connecting to the central rod of the coil any rotating device we are going to provide to the coil a circular motion into the magnetic field with a specific velocity "V".

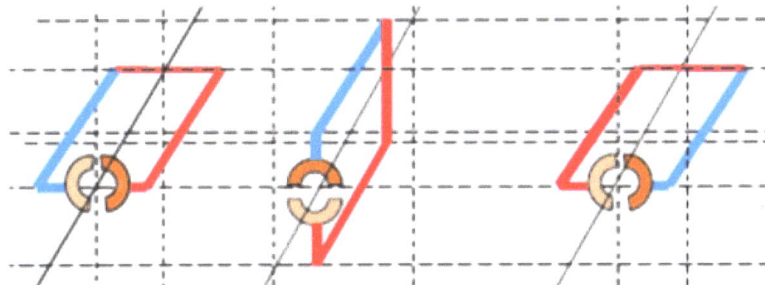

The above picture displays circular motion in three steps:

0 degree

90 degree &

180 degree

The motion continues displaying the position corresponding to:

270 degree &

360 degree

- 0 degree again

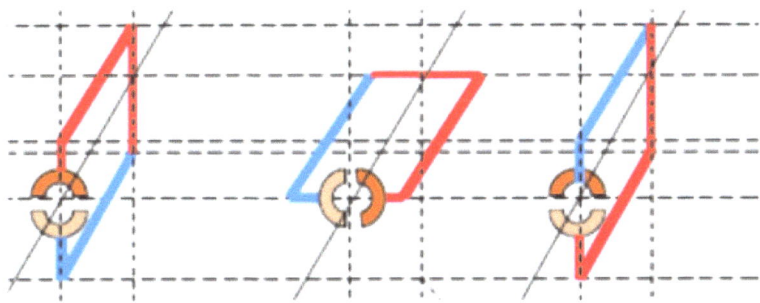

Observe the velocity vector V on the front view, just below the 3D view coil.

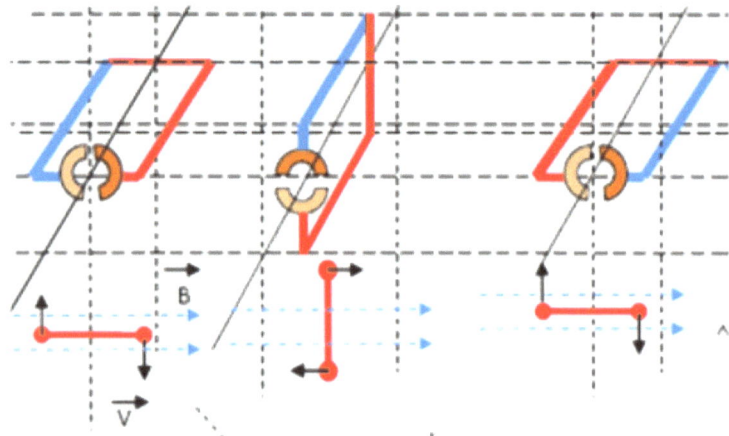

So we symbolize the motion of the coil in six positions in time: 0 degree; 90 degree; 180 degree; 270 degree; 360 degree and back to 0 degree. This is a complete rotation.

 As soon as we connect the sustaining rod of the coil directly to a coupling system and provide a rotation of the coil into the magnetic field we will be able to measure at the collectors A&B a VOLTAGE.

We agree that the coil will rotate inside of the magnetic field with constant velocity V. This in relation with intensity of the magnetic field is generating a measurable voltage at the collector's terminal A&B

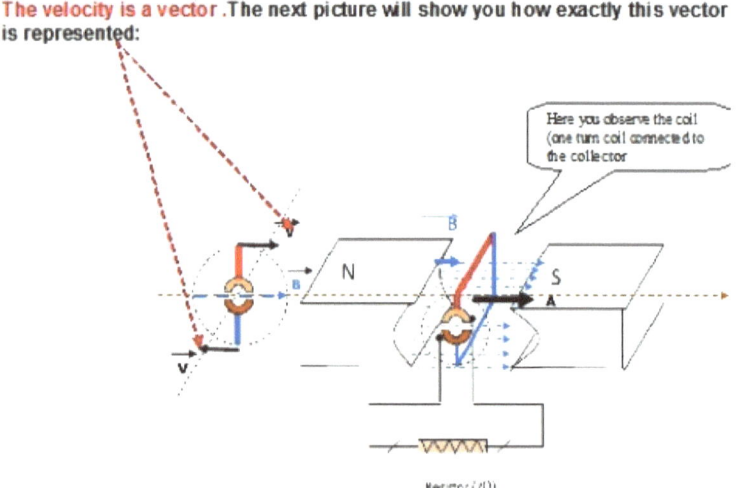

Please observe how the coil's area is cutting a maximum numbers of magnetic field lines. Since the VELOCITY is the main vector to indicate the rotation please observe the red arrow to identify the VELOCITY Vector and his direction.

In bellow this picture the VELOCITY Vector (V) is perpendicular with the magnetic field strength Vector (B)

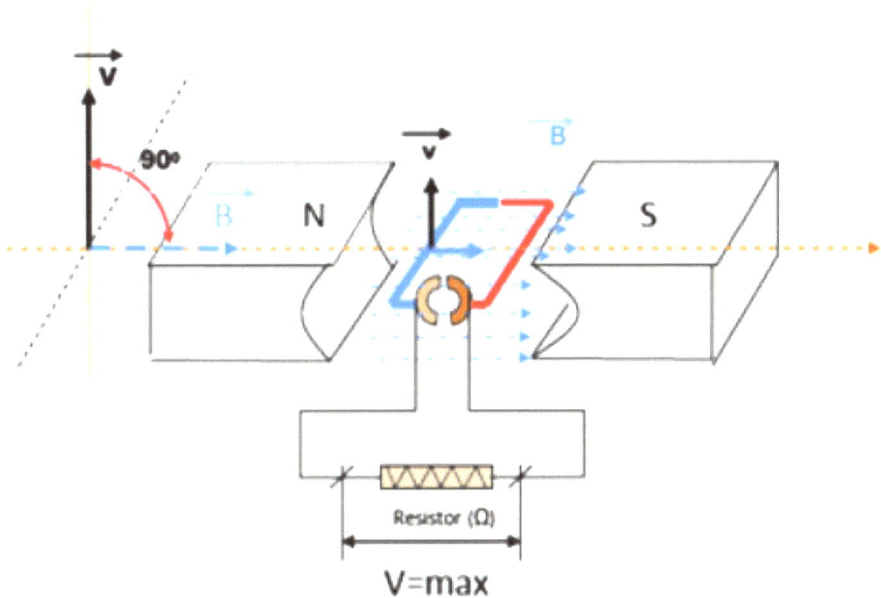

Generators

There are three types of electrical generators
- Single –phase generators

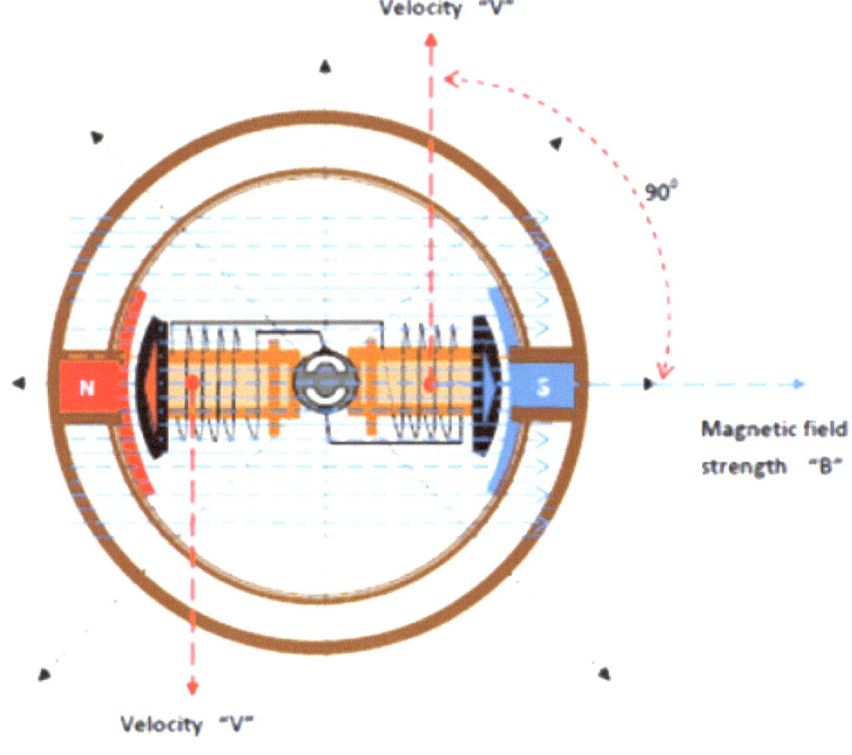

- 3 Phase generators(poly-phase generator)

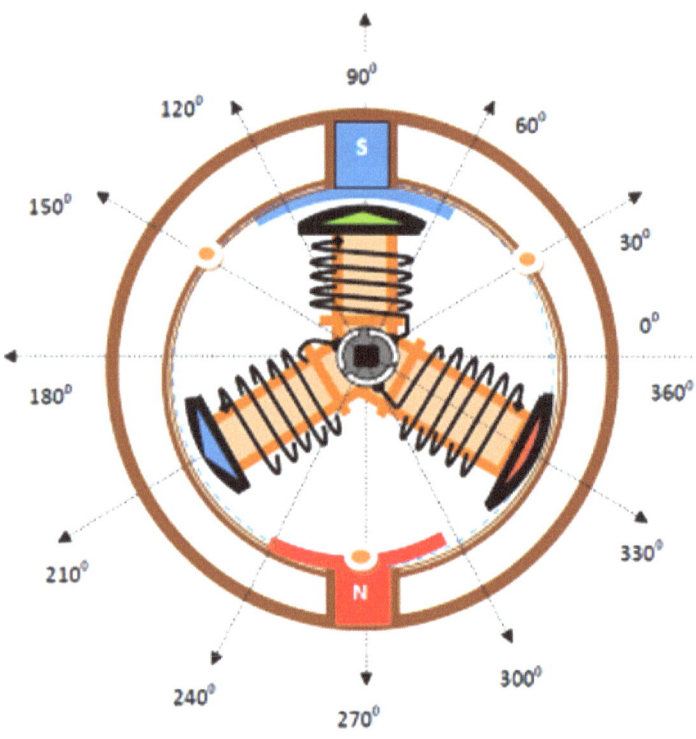

Single Phase Generator

Single –phase electrical generator is an electrical device that converts the mechanical energy to an electrical energy. This device is able to supply just one single and continuously alternating voltage at the coils of the generator. The voltage is sinusoidal function in time.
http://scratch.mit.edu/projects/13162517/#fullscreen

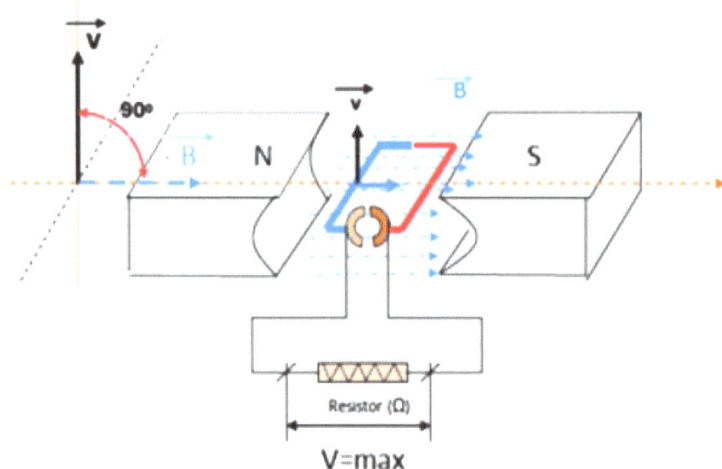

The graph below shows a typical single phase AC wavelength.

http://scratch.mit.edu/projects/13399200/#fullscreen

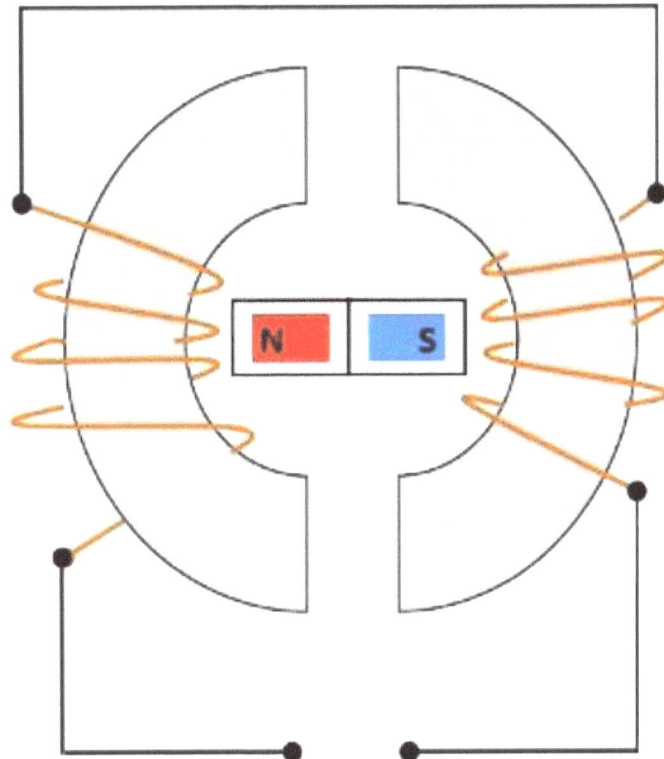

http://scratch.mit.edu/projects/13408595/#fullscreen

Generator's parts

The main important parts of generator are:

The STATOR

The ROTOR

The COLECTOR

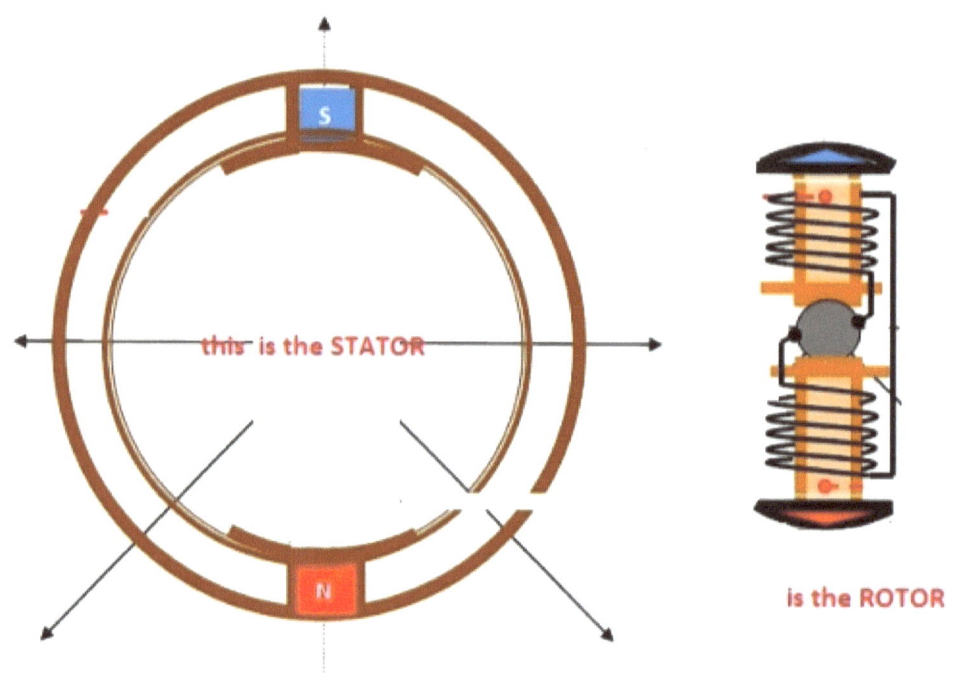

this is the STATOR

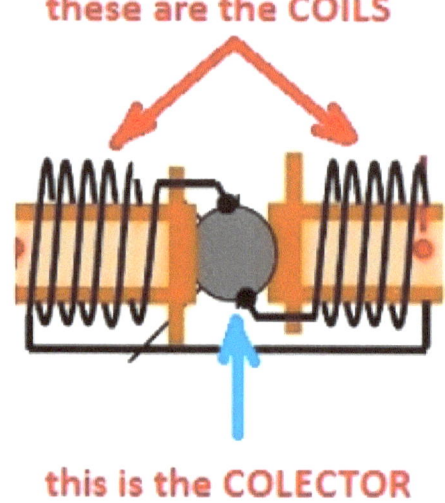

is the ROTOR

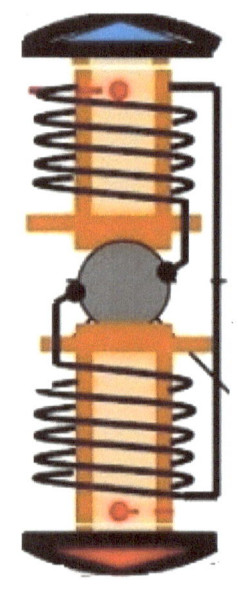

is the ROTOR

these are the COILS

this is the COLECTOR

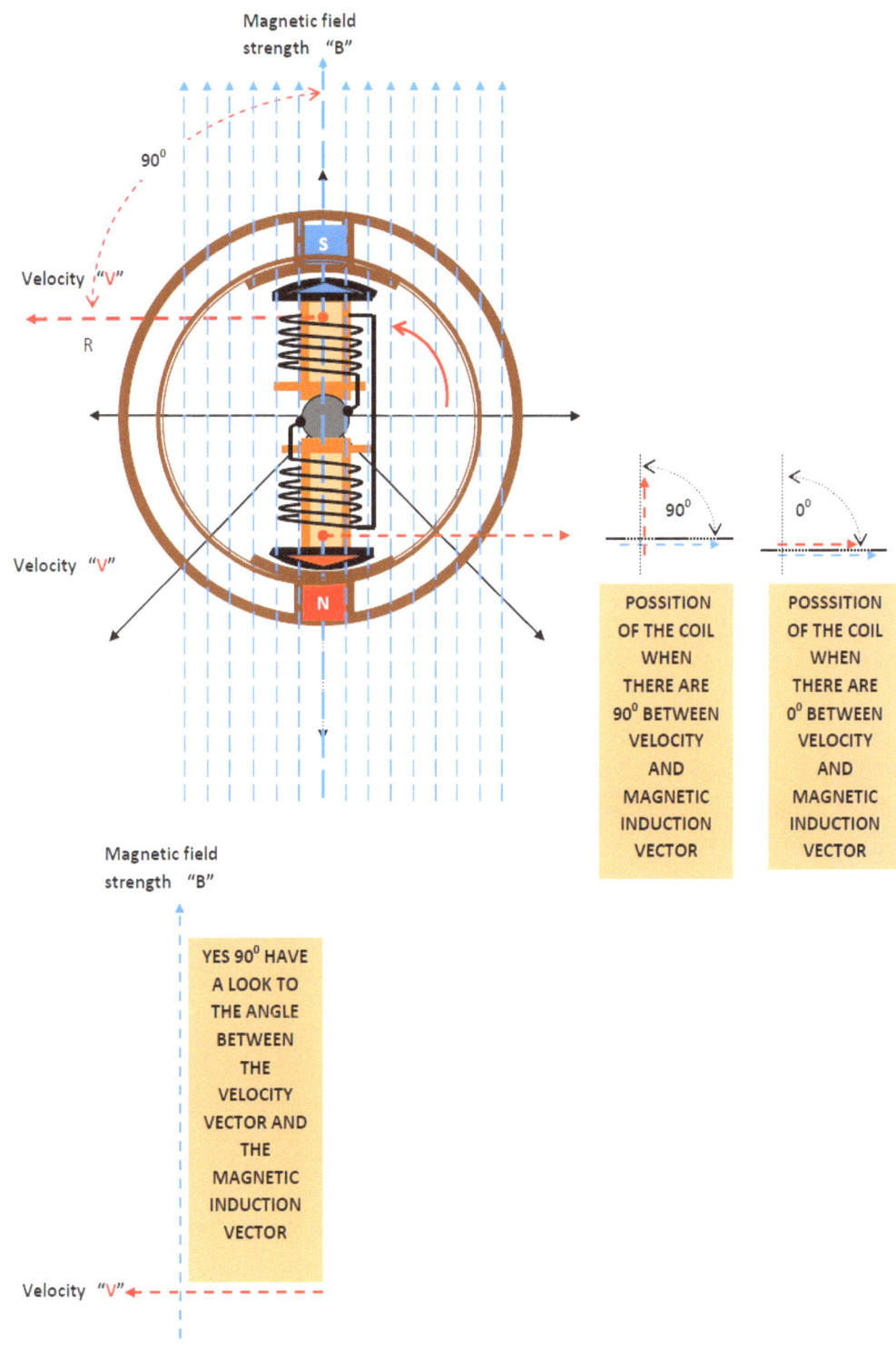

Magnetic field strength "B"

90⁰

Velocity "V"

R

Velocity "V"

Magnetic field strength "B"

Velocity "V"

90⁰

0⁰

POSSITION OF THE COIL WHEN THERE ARE 90⁰ BETWEEN VELOCITY AND MAGNETIC INDUCTION VECTOR

POSSSITION OF THE COIL WHEN THERE ARE 0⁰ BETWEEN VELOCITY AND MAGNETIC INDUCTION VECTOR

YES 90⁰ HAVE A LOOK TO THE ANGLE BETWEEN THE VELOCITY VECTOR AND THE MAGNETIC INDUCTION VECTOR

Since the VELOCITY is the main vector to indicate the rotation please observe the red arrow to identify the VELOCITY Vector and his direction.

At the position we have in this picture the VELOCITY Vector is perpendicular with the magnetic field strength Vector (B)

At this detail we see the North (N) and South(S) poles in horizontal position since the magnets are installed on the horizontal axis.

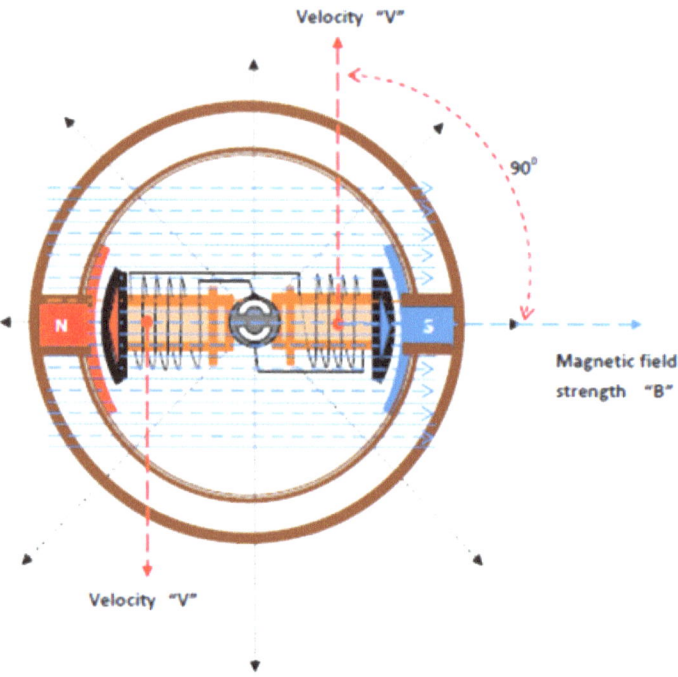

When the coil will rotate into the magnetic field we are going to identify different angles between those two main VECTORS,

- Vector V -VELOCITY
- Vector B -MAGNETIC INDUCTION

Very important is to observe the generated voltage is in direct proportional relation with:

- Angular Velocity of the coil
- Induction of the magnetic field
- Sin θ , where θ is the angle between these two vectors(V&B)
- Length of the coil "l"

The induced voltage into the coil is: **Voltage =l*V*B*Sin θ**

At this detail we see the North (N) and South(S) poles in vertical position since the magnets are installed on the vertical axis.

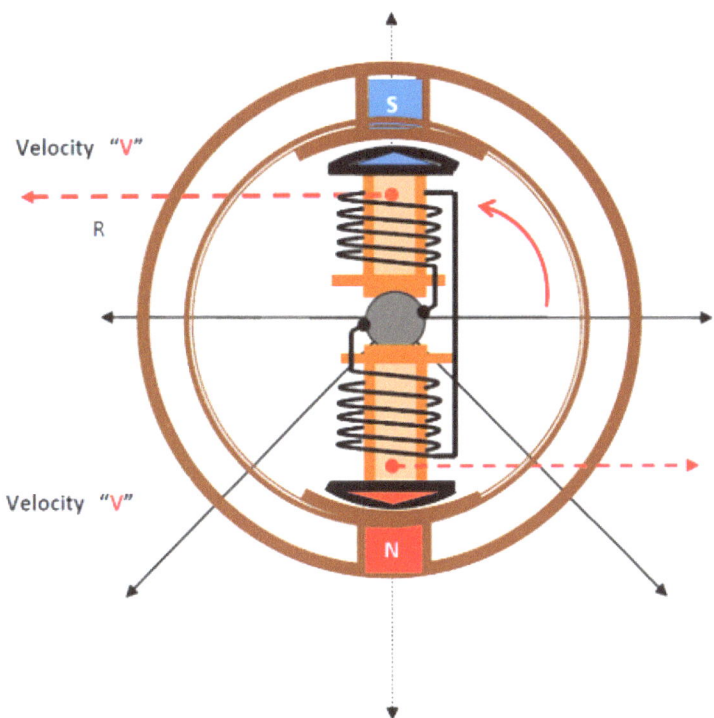

When the coil will rotate into the magnetic field we are going to identify different angles between those two main VECTORS,

- Vector V -VELOCITY
- Vector B -MAGNETIC INDUCTION

Very important is to observe the generated voltage is in direct proportional relation with:

- Angular Velocity of the coil
- Induction of the magnetic field
- Sin θ , where θ is the angle between these two vectors(V&B)
- Length of the coil "l"

The induced voltage into the coil is: **Voltage =l*V*B*Sin θ**

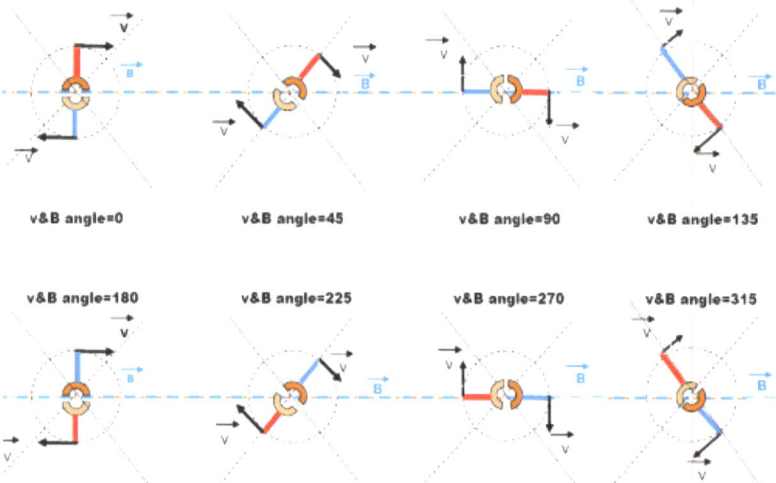

| v&B angle=0 | v&B angle=45 | v&B angle=90 | v&B angle=135 |
| v&B angle=180 | v&B angle=225 | v&B angle=270 | v&B angle=315 |

Now let assume we are going to the next step of our experiment and to apply the voltage induced into this coil to a resistor, a lamp whatever you want…

To be more explicit let see the bellow picture so we visualize such thing……

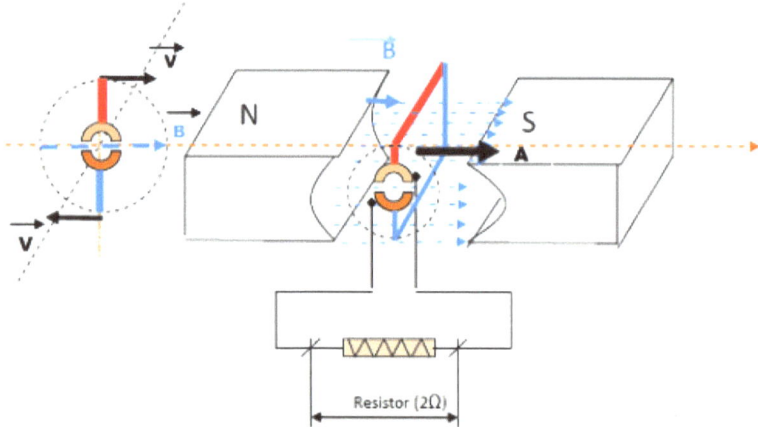

Resistor (2Ω)

Through such circuit: the resistor & the coil will flow an ELECTRIC CURRENT. The unit measure for current is the ampere. The current will pass through resistor and will generate some heat. We just discovered the voltage and current that can be used for different applications. Based on required application and voltage, current required amount we need to provide a specific generator able to generate such "product".

- The voltage will be available all time if the coil is rotating into magnetic field and will be no available if the coil stops.

- The current flows as long as the resistor is connected to the coil so the circuit is closed.

- Since the circuit is open will be no current to flow from the generator to resistor since the voltage is there.

- If coil stops will be no voltage and current to flow through resistor.

Rotating a coil into a magnetic field will generate at the terminals of the coil a voltage direct proportional with the variation of the magnetic flux in time on the area of the coil. The generated voltage will have a form of wave! In case the rotation speed is constant:
The wave is sinusoidal wave!

Voltage and Current Diagram

EFM formula is a "SIN function" of the angle between V and B .This will provide the wave shape of the EMF. When the voltage is applied to a circuit an electrical current will flow to the load.

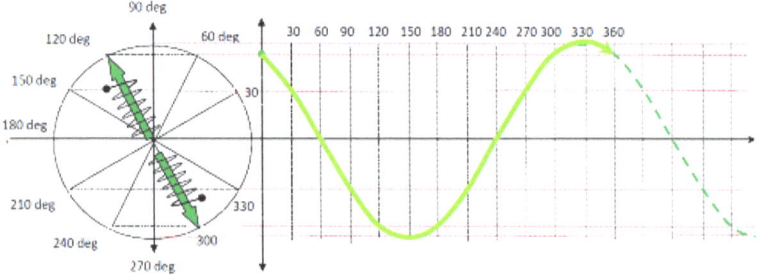

When the voltage is applied to a circuit an electrical current will flow to the load.
The load is resistive (R). The electrical current is direct proportional with the voltage applied to the circuit and inverse proportional with the resistance of the circuit. (Measured in OHM).
In such conditions the electric current's shape is going to be sin wave also.

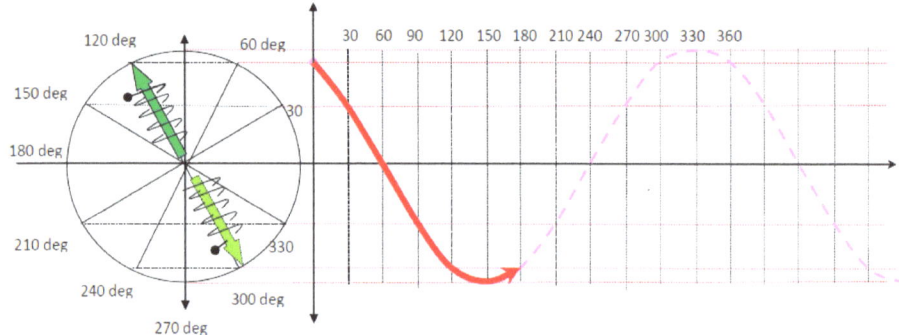

In case the load is a resistive load there the VOLTAGE and CURRENT shape is almost identical. The amplitude is different but they will touch the maximum and the minimum values at the same time. They are in phase. There is no lag (interval or delay) in between the voltage and current. Showing the Voltage and Current shape on the same system of they look like this:

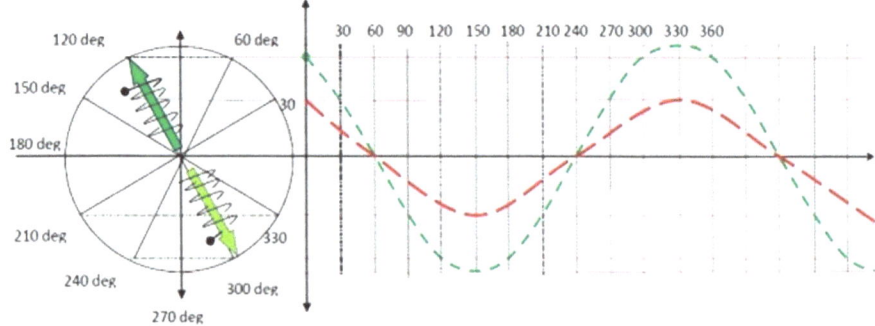

Green is the voltage since current is red shape.
I should mention again one thing you need to consider: THE CIRCUIT SI CLOSED.

Here is represented the moment when the V& B vectors are perpendicular relative to each other.

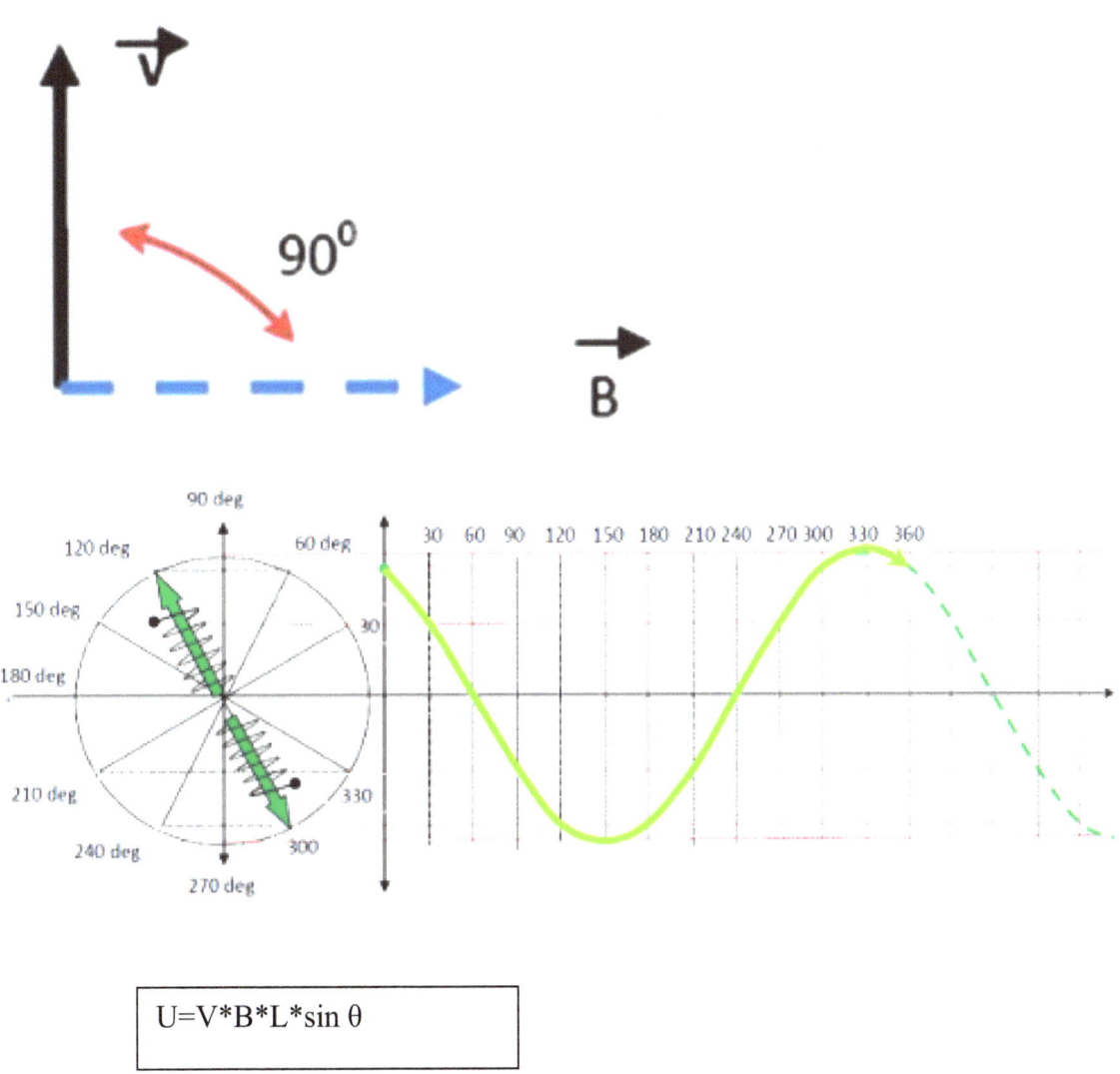

$$U = V*B*L*\sin\theta$$

Why sine wave?

http://scratch.mit.edu/projects/13162517/#fullscreen

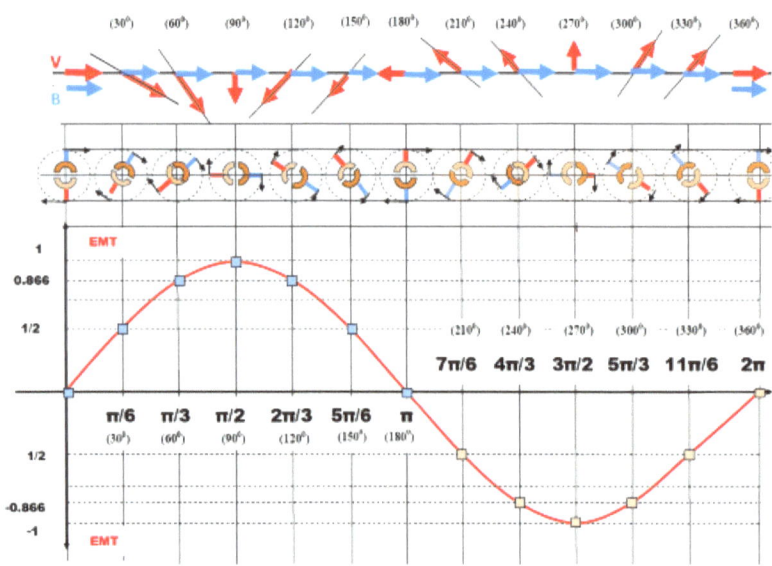

Because the voltage is depending of **sin θ**

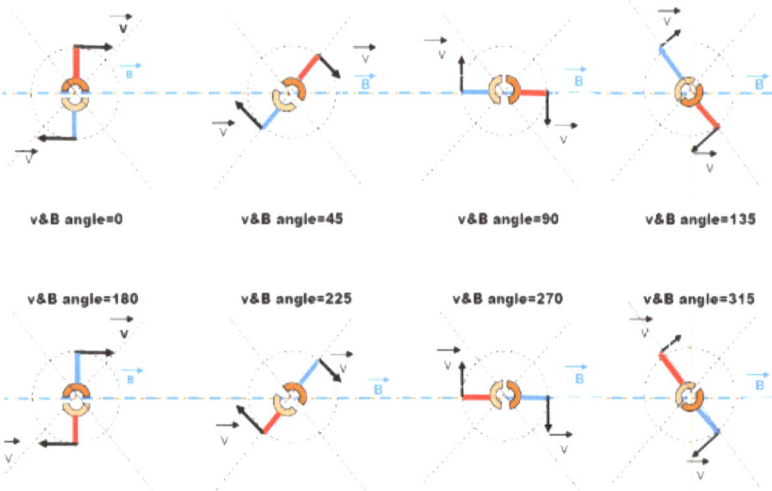

http://scratch.mit.edu/projects/13399200/#fullscreen

http://scratch.mit.edu/projects/13852283/#fullscreen

We are going to study, how and why the form of the EMF and current has such shape for the single-phase electrical generator.

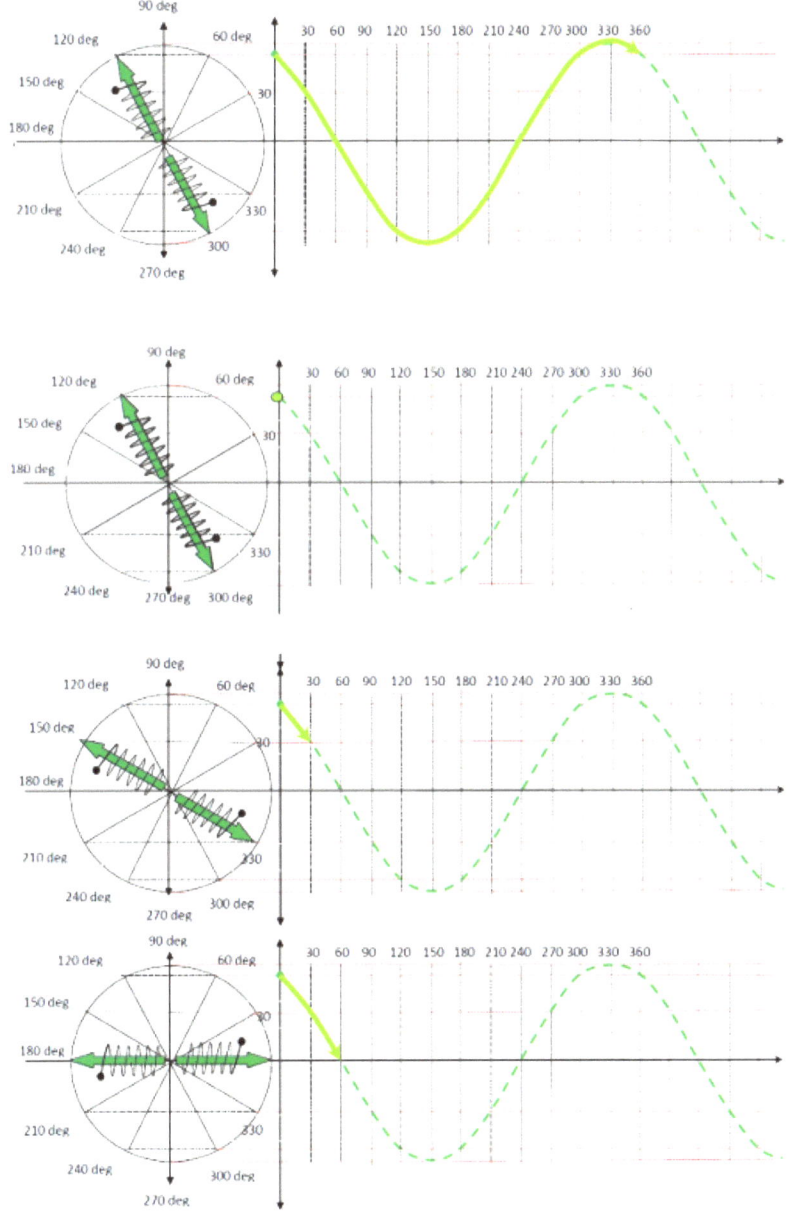

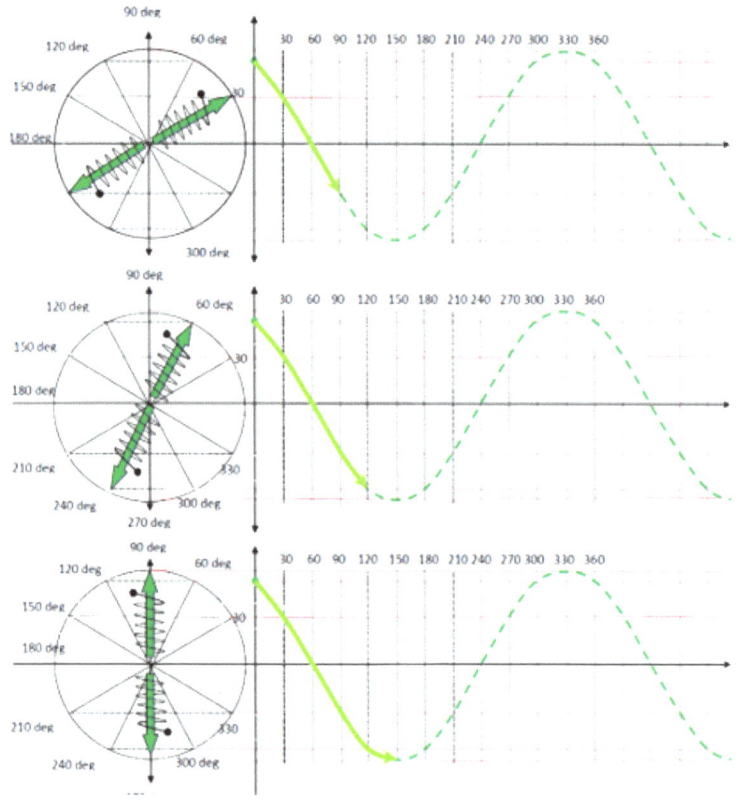

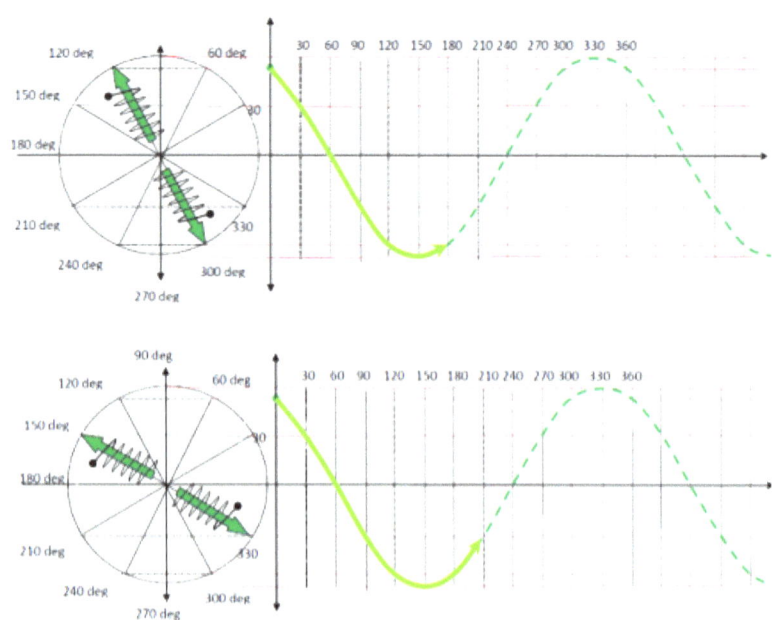

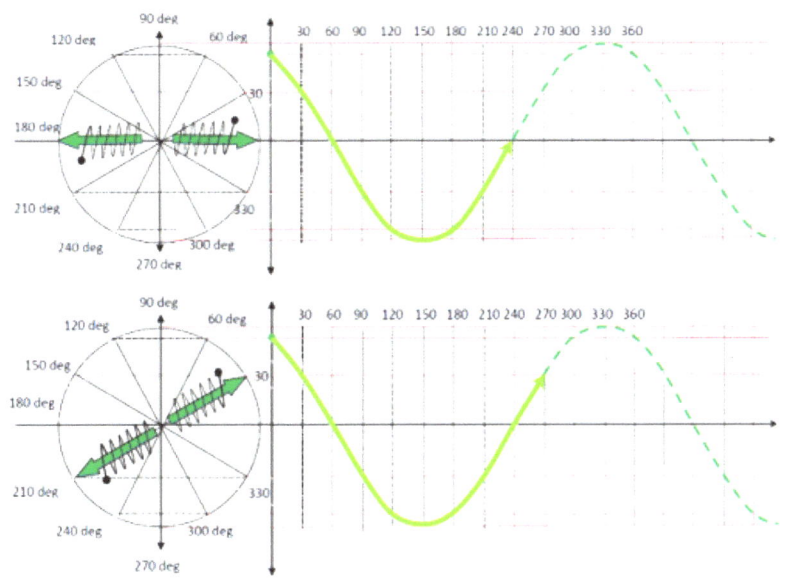

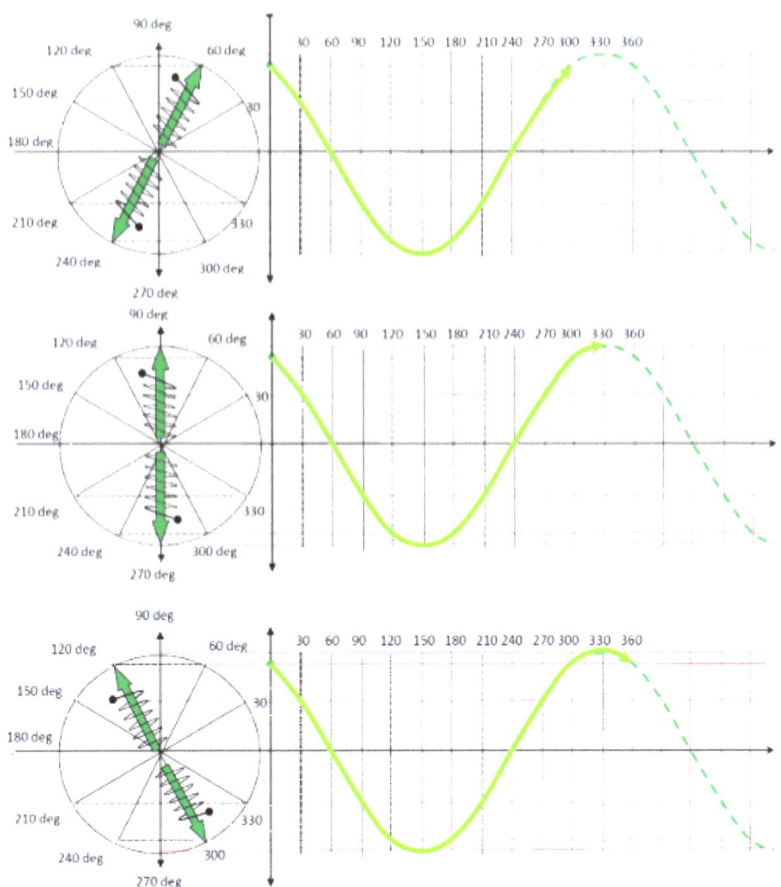

Three Phase Generator

http://scratch.mit.edu/projects/12802756/#fullscreen

Three-phase electrical generator is an electrical device that converts the mechanical energy to an electrical energy. This device is able to supply three continuously alternating voltages at the coils of the generator. The voltage for each coil is sinusoidal function in time but starts at different moments for each coil.The graph below shows a typical single phase AC wavelength.

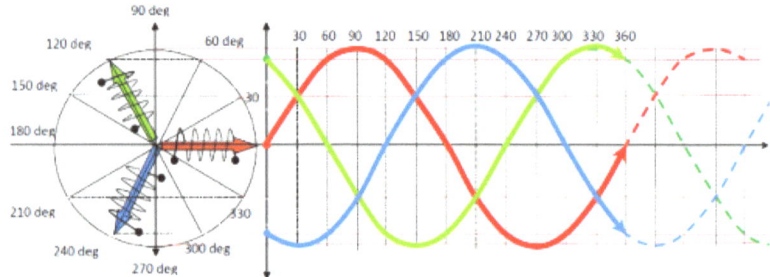

Generator's parts

The main important parts of generator are:
The STATOR

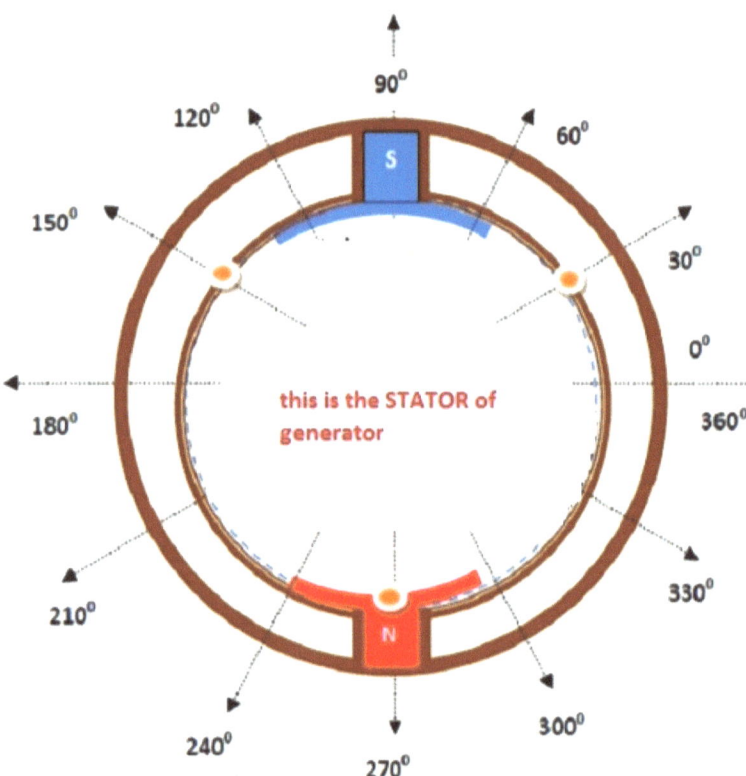

The ROTOR
http://scratch.mit.edu/projects/12778825/

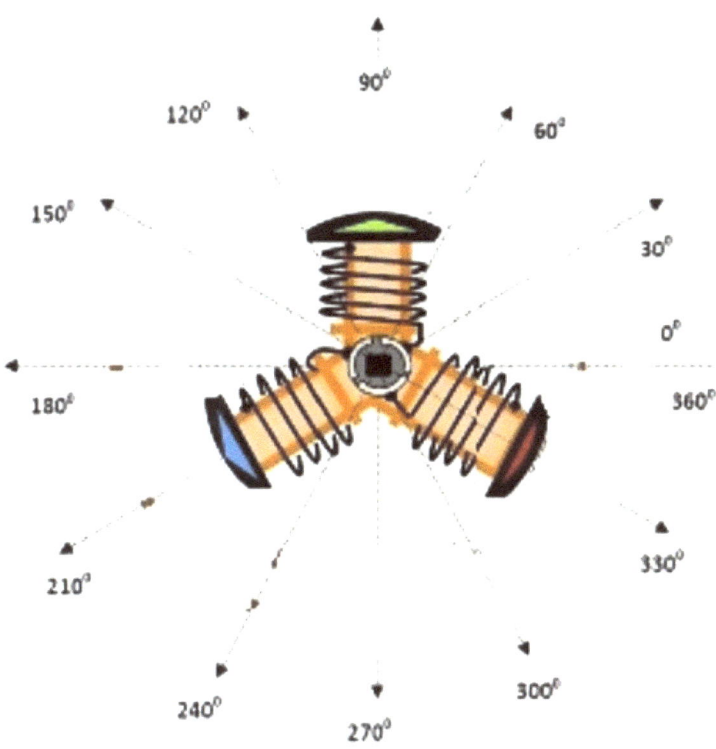

The COLECTOR

These are the COILS(3 ea)

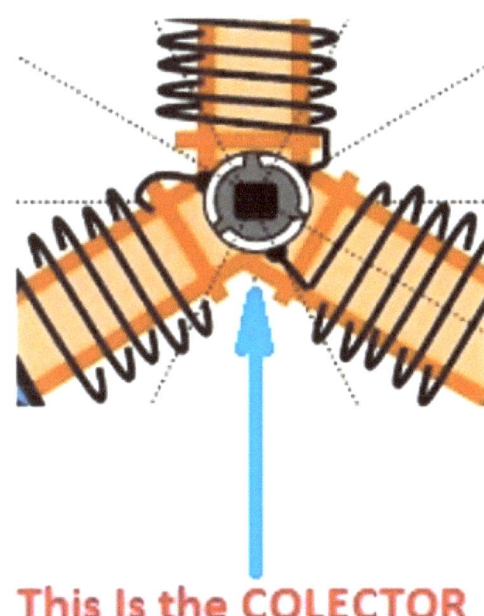

This Is the COLECTOR

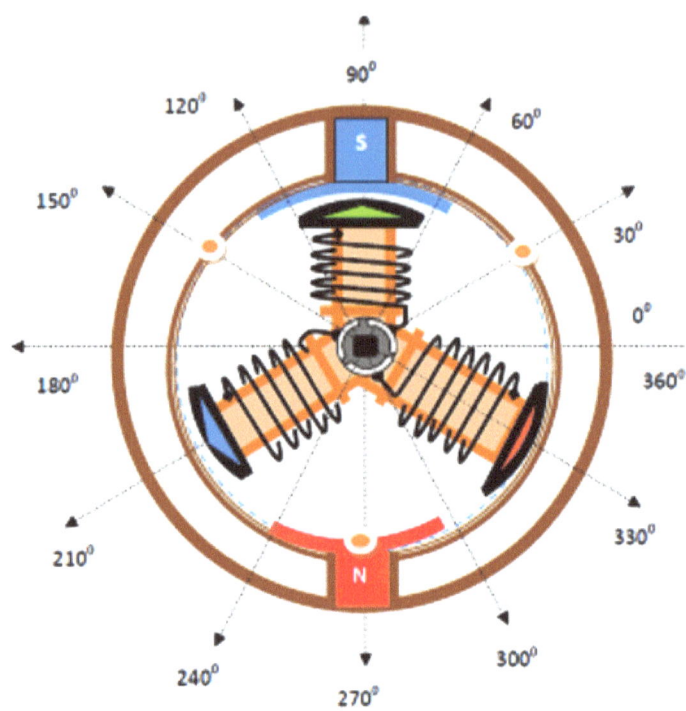

Since the VELOCITY is the main vector to indicate the rotation please observe the red arrow to identify the VELOCITY Vector and his direction.

At the position we have in this picture the VELOCITY Vector is perpendicular with the magnetic field strength Vector (B)

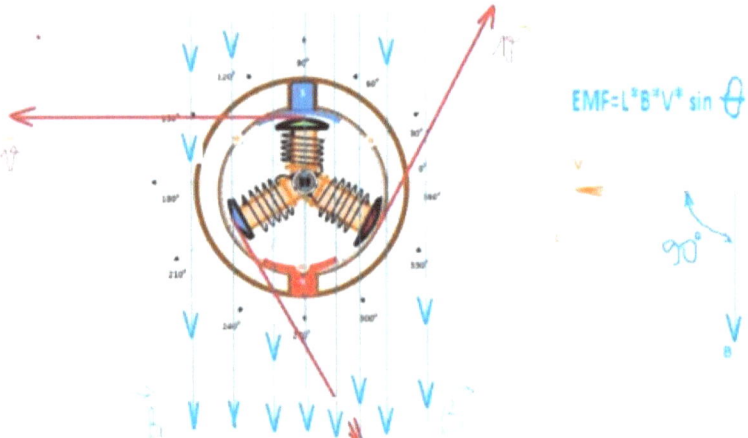

At this detail we see the North (N) and South(S) poles in horizontal position since the magnets are installed on the horizontal axis.

When the coil will rotate into the magnetic field we are going to identify different angles between those two main VECTORS,

- Vector V -VELOCITY
- Vector B -MAGNETIC INDUCTION

Very important is to observe the generated voltage is in direct proportional relation with:

- Angular Velocity of the coil
- Induction of the magnetic field
- Sin θ , where θ is the angle between these two vectors(V&B)
- Length of the coil "l"

The induced voltage into the coil is: **Voltage =l*V*B*Sin θ**

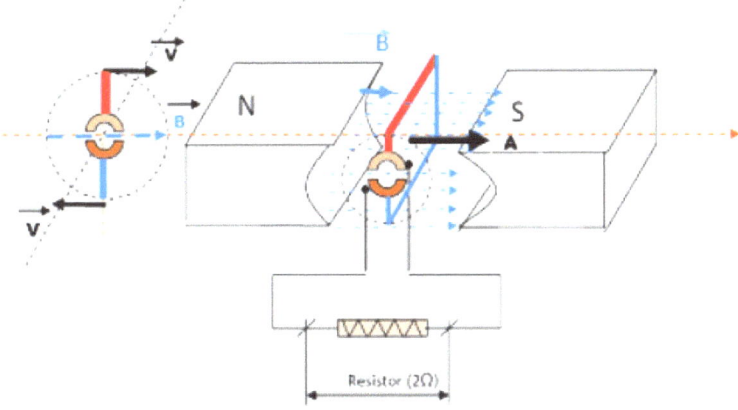

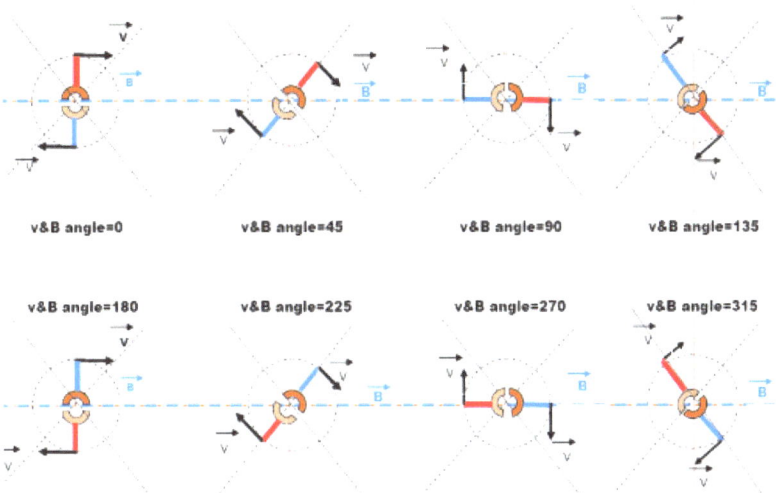

| v&B angle=0 | v&B angle=45 | v&B angle=90 | v&B angle=135 |

| v&B angle=180 | v&B angle=225 | v&B angle=270 | v&B angle=315 |

Now let assume we are going to the next step of our experiment and to apply the voltage induced into this coil to a resistor, a lamp whatever you want…

To be more explicit let see the bellow picture so we visualize such thing......

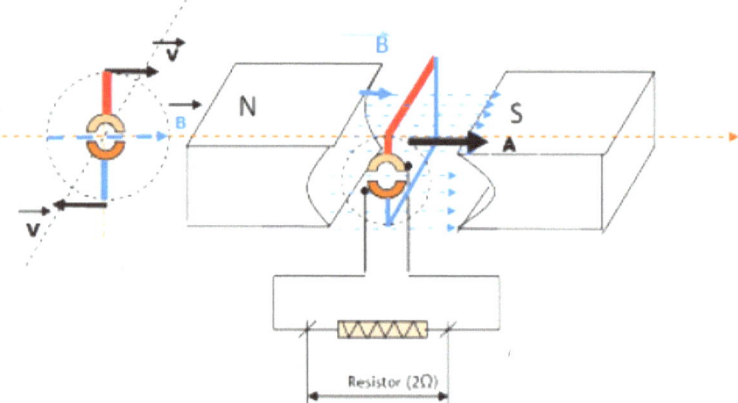

Through such circuit: the resistor & the coil will flow an ELECTRIC CURRENT. The unit measure for current is the ampere. The current will pass through resistor and will generate some heat. We just discovered the voltage and current that can be used for different applications.

Based on required application and voltage, current required amount we need to provide a specific generator able to generate such "product".

•   The voltage will be available all time if the coil is rotating into magnetic field and will be no available if the coil stops.

•   The current flows as long as the resistor is connected to the coil so the circuit is closed.

•   Since the circuit is open will be no current to flow from the generator to resistor since the voltage is there.

•   If coil stops will be no voltage and current to flow through resistor.

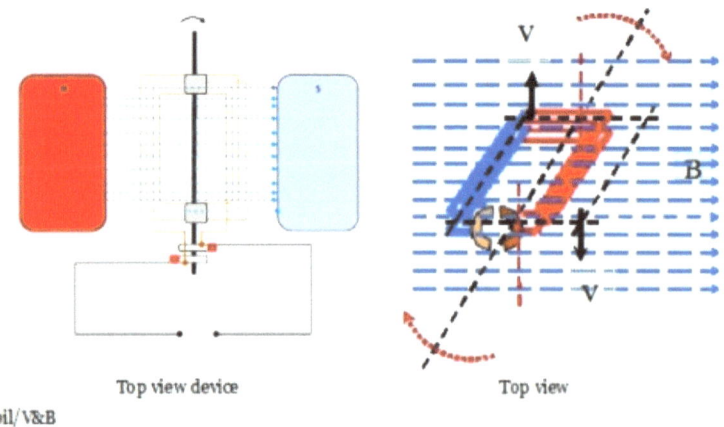

Top view device   Top view

coil/ V&B

Rotating a coil into a magnetic field will generate at the terminals of the coil a voltage direct proportional with the variation of the magnetic flux in time on the area of the coil. The generated voltage will have a form of wave! In case the rotation speed is constant:
The wave is sinusoidal wave!

The Stator

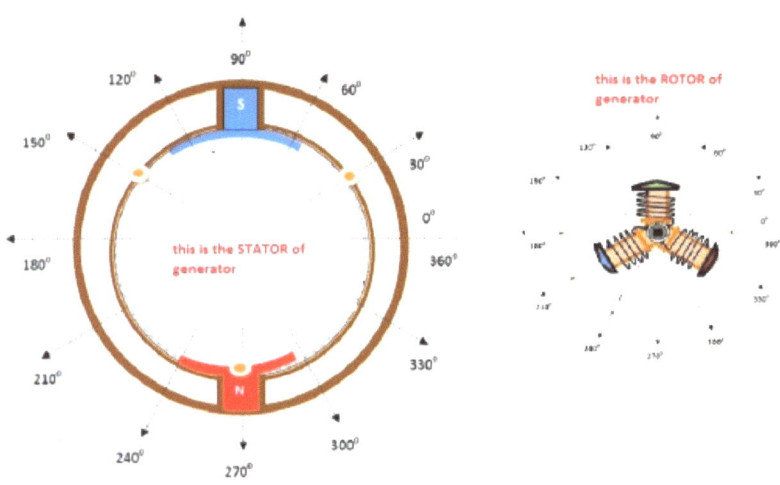

The Rotor

http://scratch.mit.edu/projects/12778825/

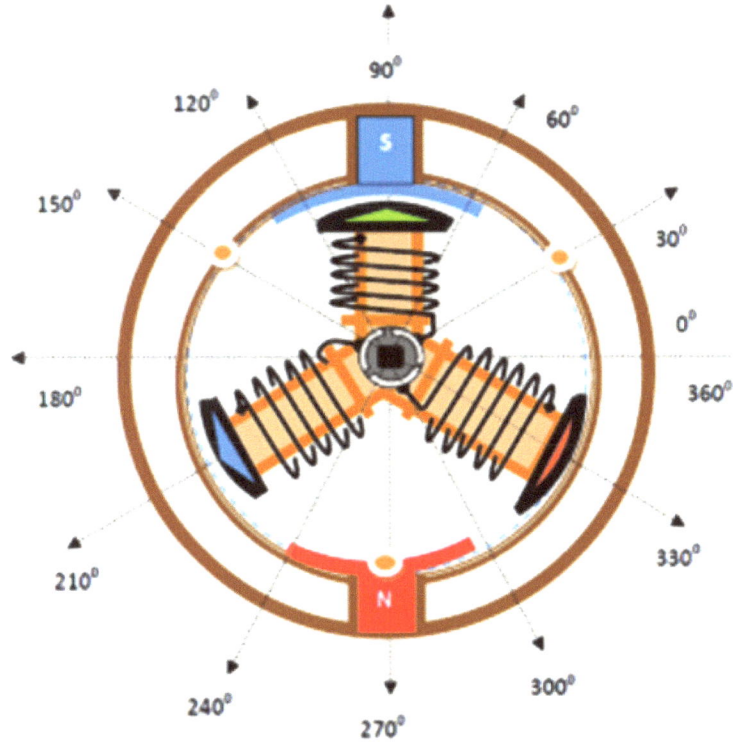

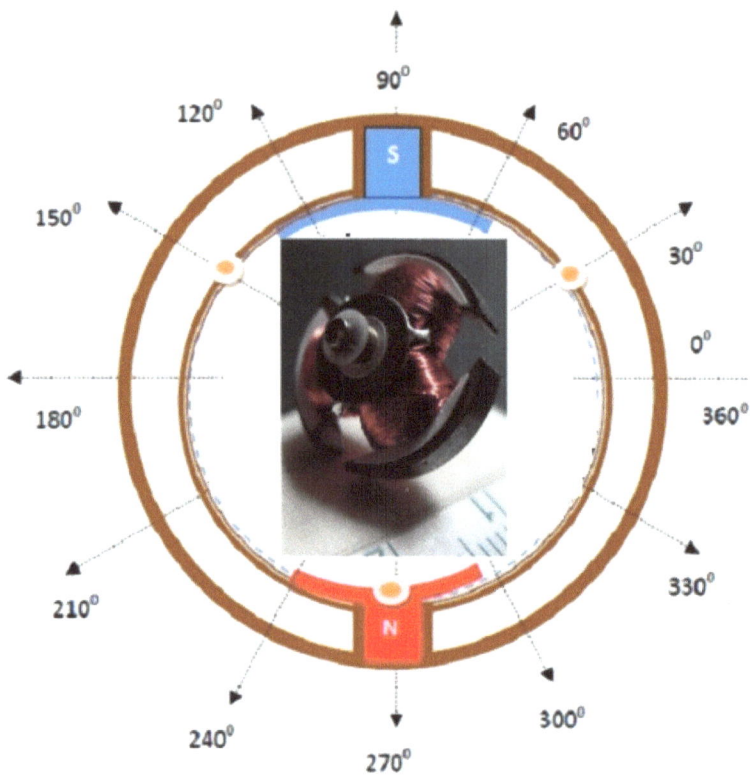

The Collector

Voltage and Current Diagram

EFM formula is a "SIN function" of the angle between V and B .This will provide the wave shape of the EMF. When the voltage is applied to a circuit an electrical current will flow to the load.

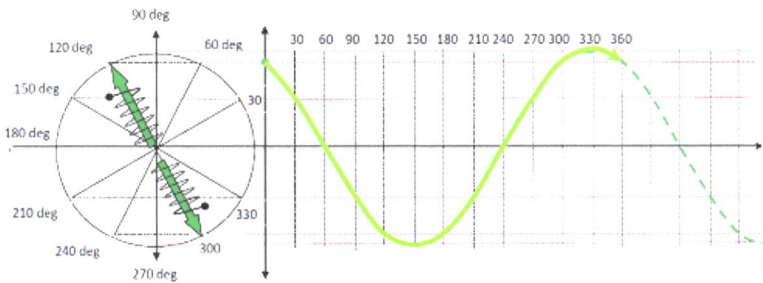

When the voltage is applied to a circuit an electrical current will flow to the load.
The load is resistive (R). The electrical current is direct proportional with the voltage applied to the circuit and inverse proportional with the resistance of the circuit. (Measured in OHM).
In such conditions the electric current's shape is going to be sin wave also.

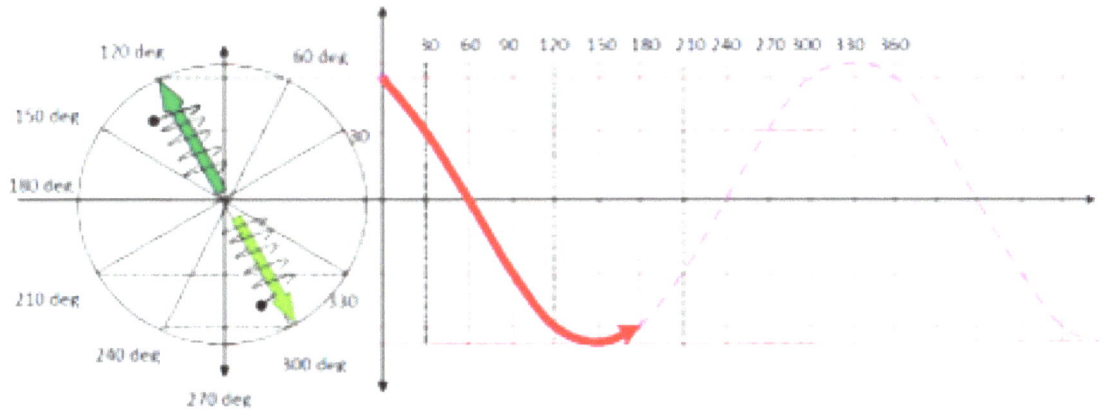

In case the load is a resistive load there the VOLTAGE and CURRENT shape is almost identical. The amplitude is different but they will touch the maximum and the minimum values at the same time. They are in phase. There is no lag (interval or delay) in between the voltage and current. Showing the Voltage and Current shape on the same system of they look like this:

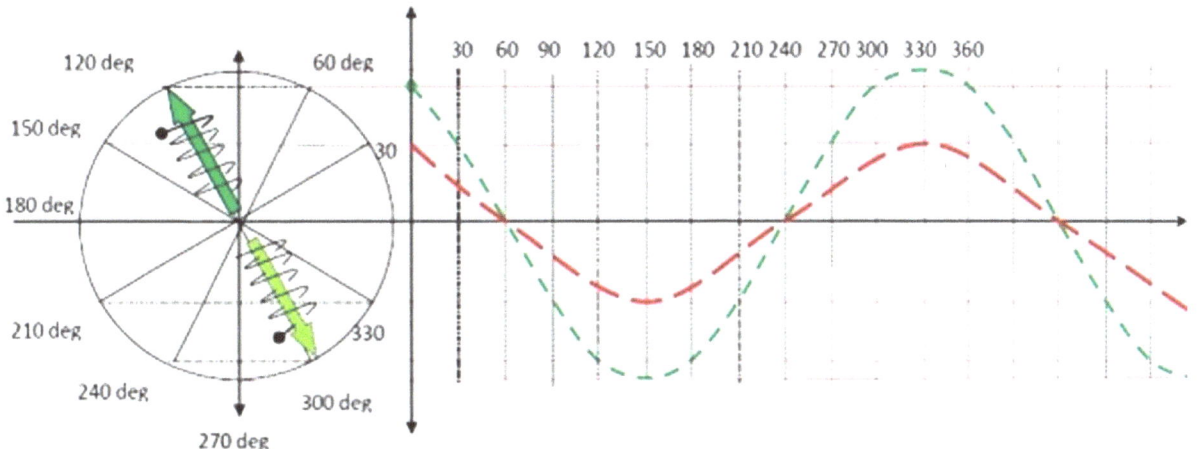

Green is the voltage since current is red shape.
I should mention again one thing you need to consider: THE CIRCUIT SI CLOSED.

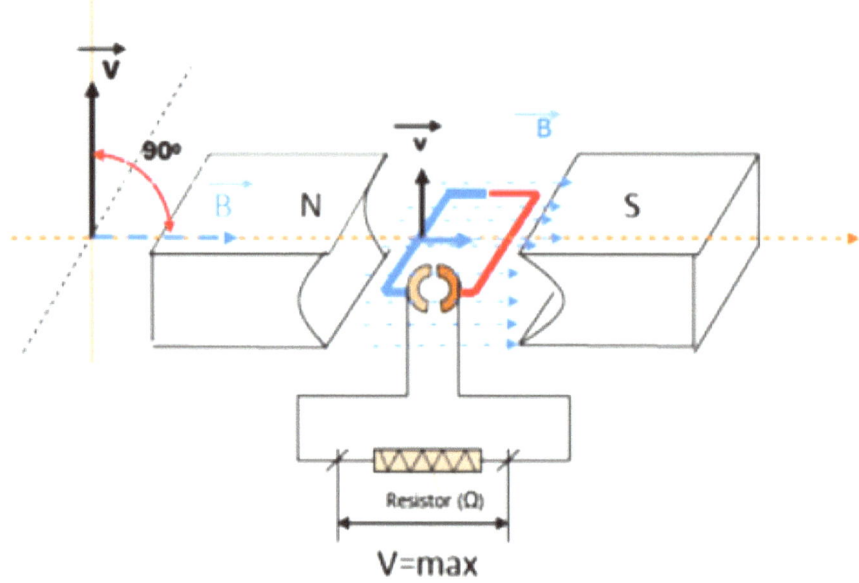

Here is represented the moment when the V& B vectors are perpendicular relative to each other.

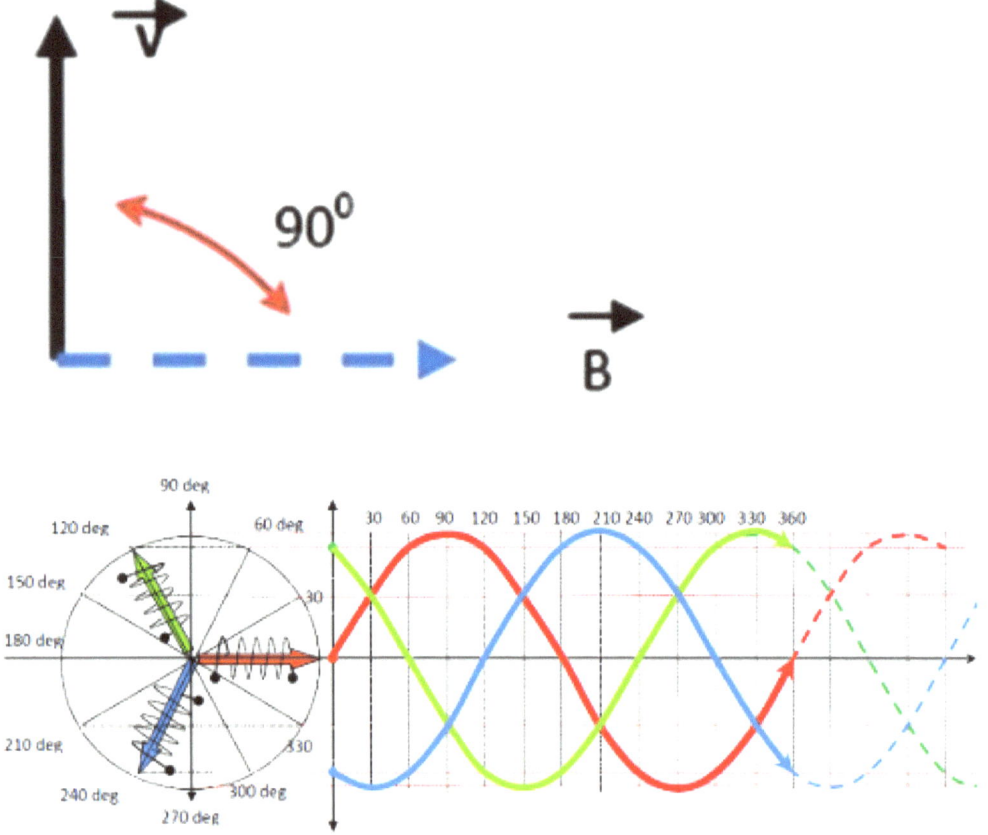

Why sine wave?

We are going to study, how and why the form of the EMF and current has such shape for each coil of the three-phase electrical generator.

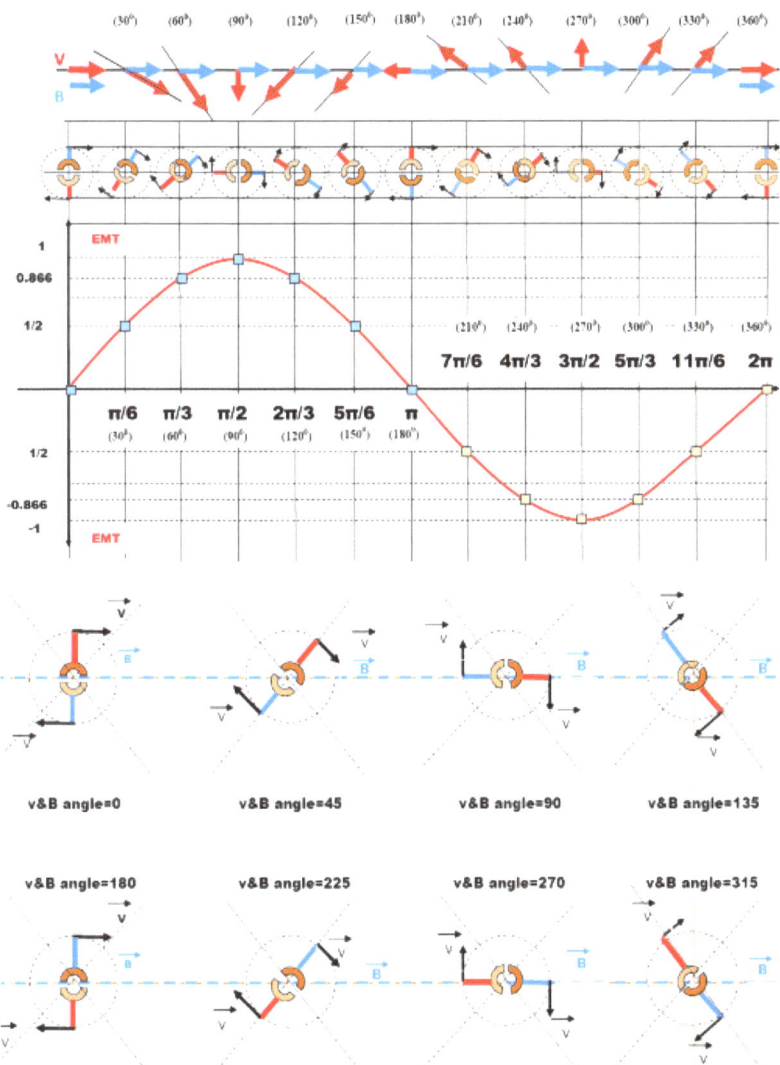

PHASE 1

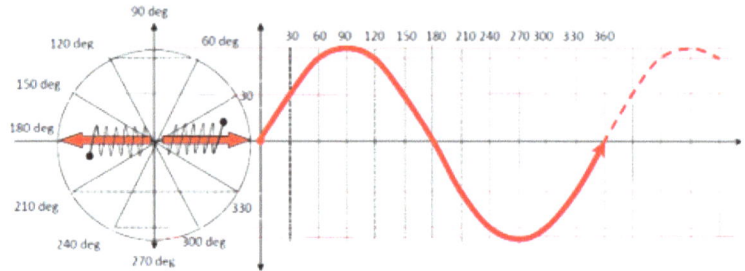

http://scratch.mit.edu/projects/13399910/#fullscreen

PHASE 2

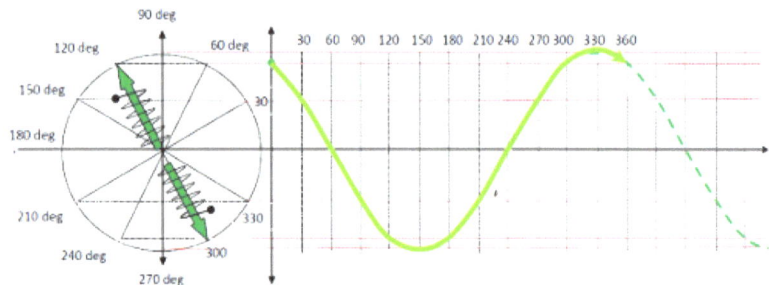

http://scratch.mit.edu/projects/13399936/#fullscreen

PHASE 3

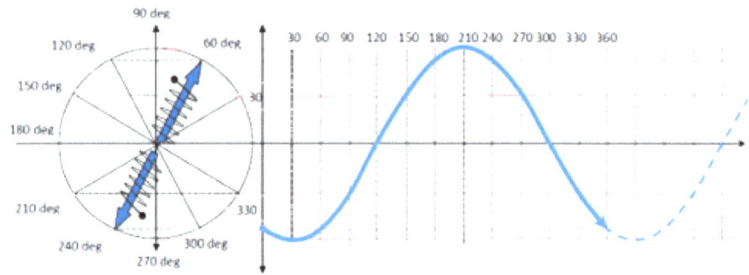

http://scratch.mit.edu/projects/13399997/#fullscreen

PHASE 1;2;3

http://scratch.mit.edu/projects/12802756/#fullscreen

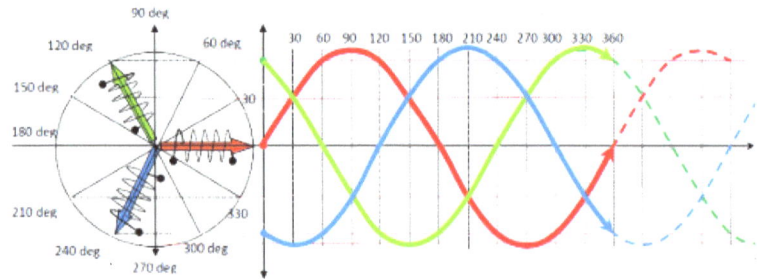

Let see the impact of RED COIL rotation step by step from zero to 360 degree .30 degree each step. GREN and BLUE coil s will rotate also. Observe how the sine wave is progressing…

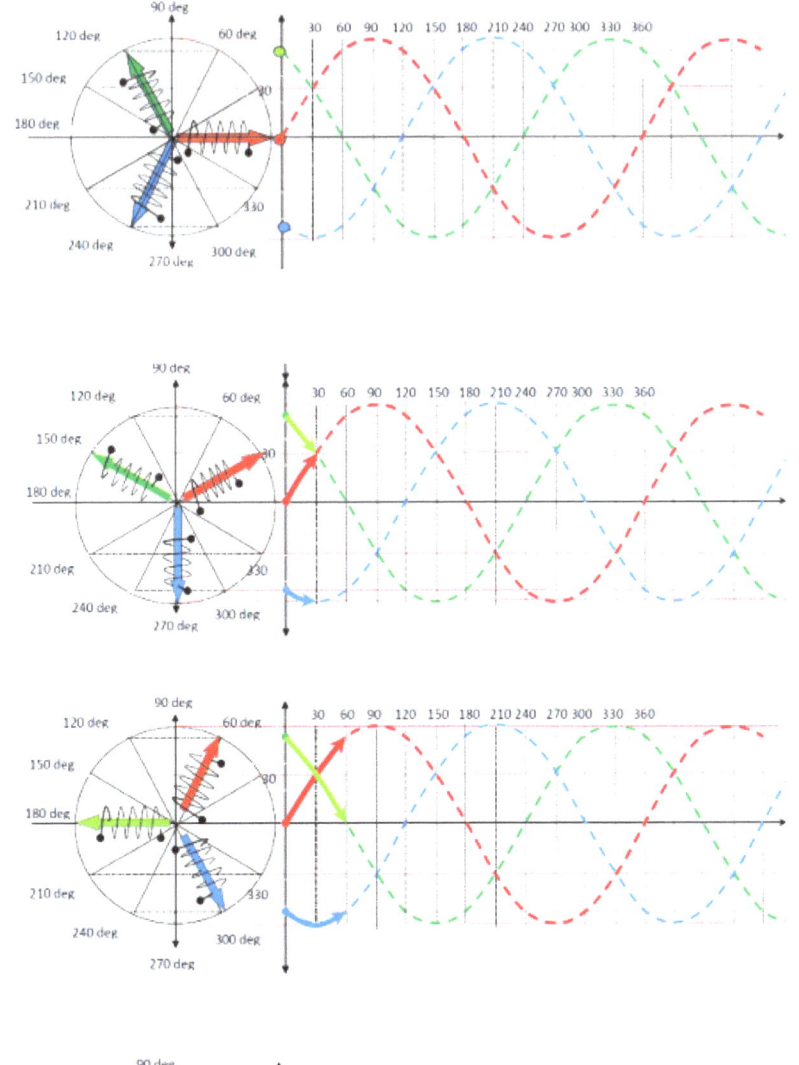

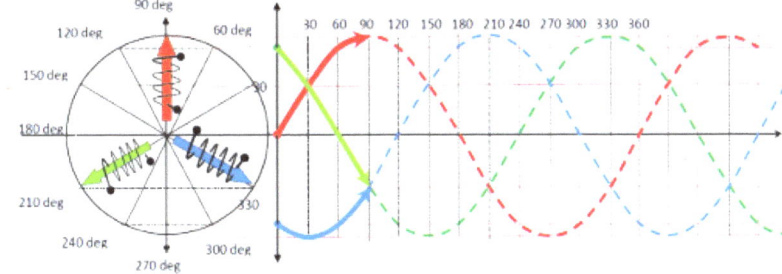

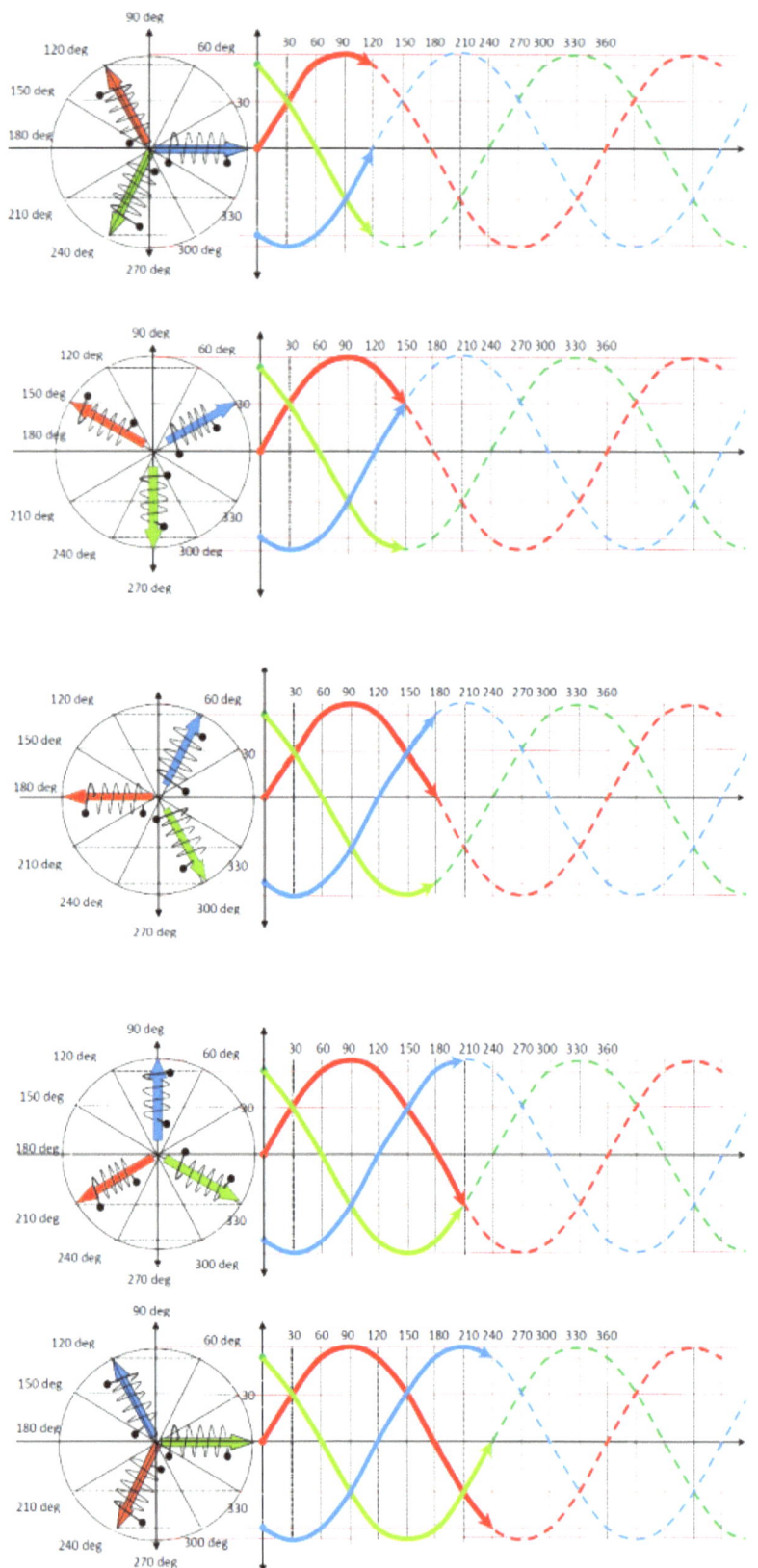

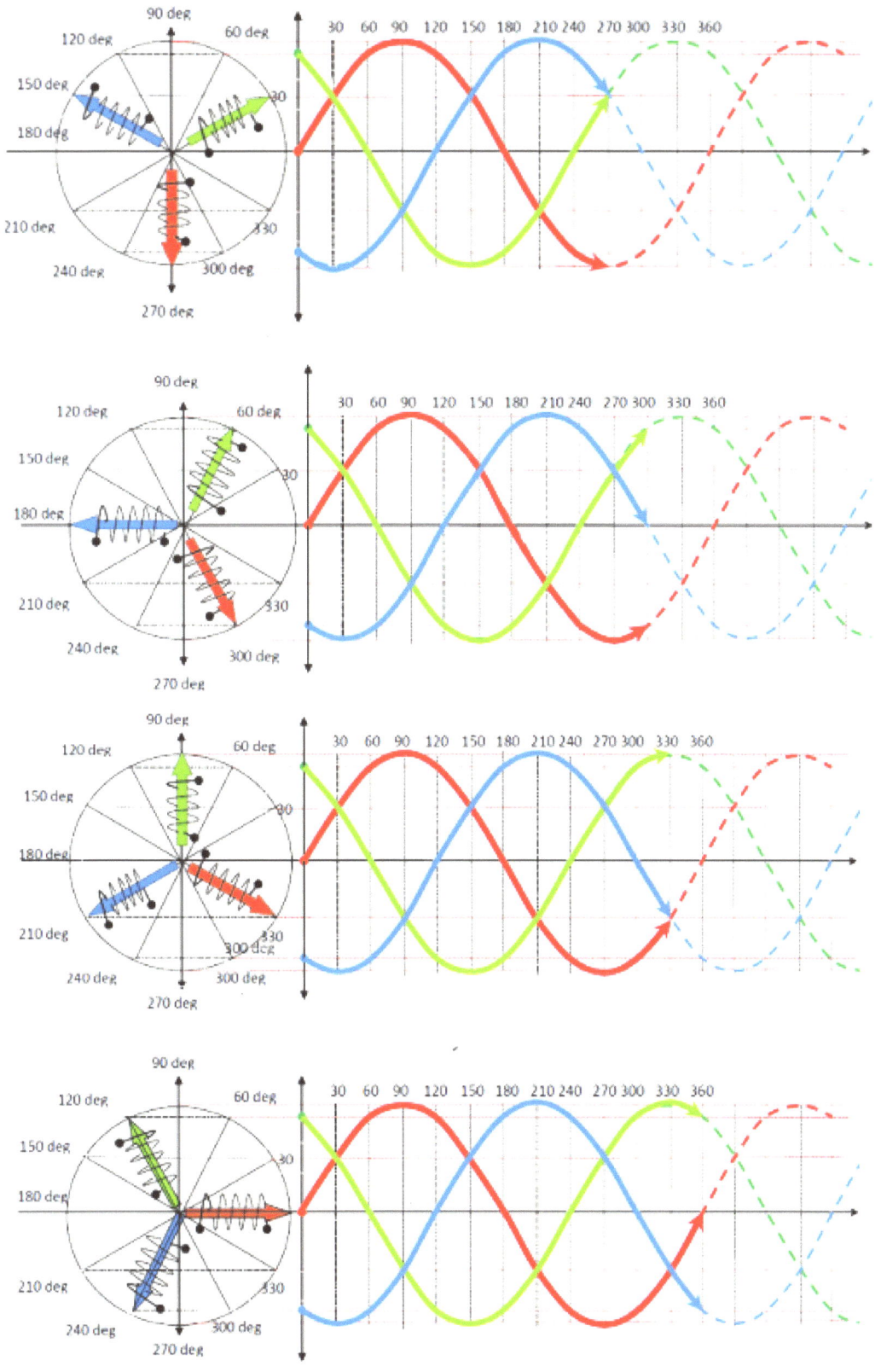

$$\Phi = B * A * Cos\ \Theta$$

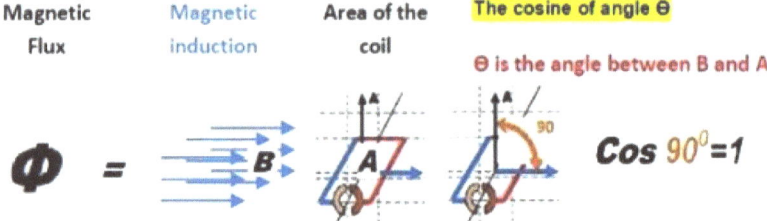

Magnetic Flux | Magnetic induction | Area of the coil | The cosine of angle Θ

Θ is the angle between B and A

$$\Phi = \quad B \quad A \quad Cos\ 90^0 = 1$$

Back Cover

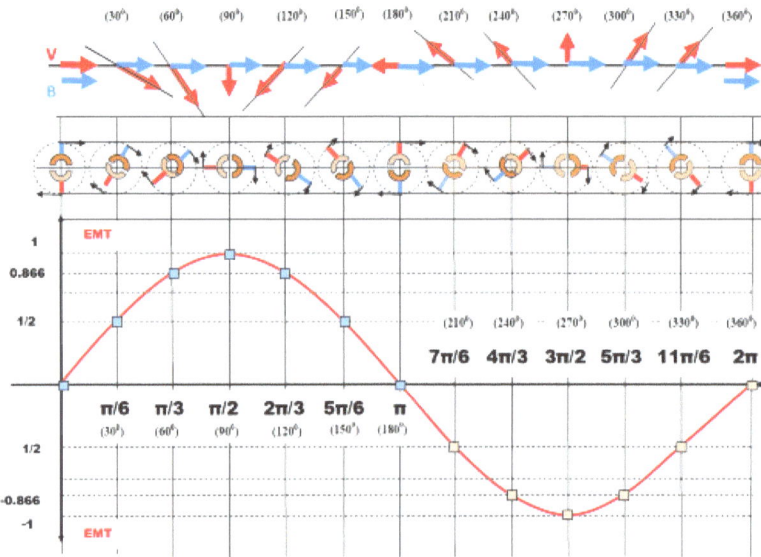

For any question please e-mail me at:

axaelectric@yahoo.ca

Thank you for reading my book, and please take a minute and make your review. In case I disappointed you I will consider your point of view and make a revision for this book. Your review is important for me!

Best regards,
Cornel Barbu